Timea Gerczei, Scott Pattison
Biochemistry Laboratory Manual For Undergraduates
An Inquiry-Based Approach

Timea Gerczei, Scott Pattison

Biochemistry Laboratory Manual for Undergraduates

An Inquiry-Based Approach

Managing Editor: Anna Rulka

DE GRUYTER
OPEN

Published by De Gruyter Open Ltd, Warsaw/Berlin
Part of Walter de Gruyter GmbH, Berlin/Munich/Boston

ISBN: 978-3-11-041132-4
e-ISBN: 978-3-11-041133-1

Bibliographic information published by the Deutsche Nationalbibliothek The Deutsche Nationalbibliothek lists this publication in the Deutsche Nationalbibliografie; detailed bibliographic data are available in the Internet at http://dnb.dnb.de.

Managing Editor: Anna Rulka

www.degruyteropen.com

Cover illustration: © Thinkstock / agsandrew

Preface

The purpose of this laboratory manual is to introduce undergraduate students to techniques used in biochemistry and molecular biology laboratories and ensure that they master the lab skills necessary to be competitive in the job market. We present a collection of fifteen experiments that teach students sterile techniques, accurate pipetting, centrifuge usage, PCR, DNA purification, protein expression and purification, HPLC, enzyme kinetics, equilibrium binding assays and introduction to bioinformatics. The emergence of bioinformatics is one of the biggest change that happened to biochemistry in the last two decades. The availability of genome sequences increased exponentially, online data banks and free programs are now available to make sense of these data. As a result we can learn about a biomolecule before ever lifting a pipette in the lab. These resources are invaluable to today's biochemists when they set up a working hypothesis. We expect a continued increase in the availability of data mining programs that help interpret the tremendous amount of genome sequence, structure, microarray etc. data thus preparing 21st century biochemists to use these programs is a must. To our knowledge, this is the only manual that includes several chapters on the latest advancements in bioinformatics: how to access genome databank, perform sequence alignments, design primers, to predict secondary and tertiary structure and to use protein visualization tools.

The unique feature of this laboratory manual is a hypothesis-driven real-life research project. In this project, students study how a noncoding RNA molecule that plays an important role in bacterial antibiotic resistance recognizes its target antibiotic. By including a research project in the undergraduate lab, students learn how real-life research works: first they set up a hypothesis then design experiments to test the hypothesis and finally evaluate the hypothesis using a functional study. With this experience students get as close to doing a research project as possible within the framework of an undergraduate laboratory. During the nine-weeks project incorporated into this laboratory manual students perform sequence alignment to determine evolutionary conserved elements in the noncoding RNA, they design primers to make mutants using site-directed mutagenesis then synthesize and purify the noncoding RNA mutants. Finally, they test the ability of the mutants to recognize a target antibiotic using a fluorescence-based binding assay. A big challenge in teaching upper level labs is the very different background and experience level students come to the course with. Some students have no biochemistry lab experience while others have done research as an undergraduate for years. Since students test their own mutant design even the most experienced students remain engaged with the process while the less experienced ones get their first taste of biochemistry research. The design of the mini project is flexible: experiments may be done in a different order or on a different target. At the authors institution the nine-week long research project was taught in the order listed in the manual: mutant design, synthesis and functional study. Alternatively, the order may be reversed: students start with analyzing a previously made noncoding RNA mutant and based on their findings they design a better mutant. A collection of ykkCD RNA mutants is available from the authors' upon request. As written, the mini project is designed to understand how the ykkCD noncoding RNA recognizes its target antibiotic tetracycline, but the project with minimal modifications can be used to examine any biologically important noncoding RNA, such as a ribozyme or regulator.

A biochemistry laboratory course is required for chemistry majors, biochemistry majors and may be chosen as an elective by pre-health professionals or students focusing on the molecular aspects of biology. This laboratory manual takes advantage of the fact that a biochemistry lecture course is pre- or co-requisite to taking a biochemistry lab course. As a result students are expected to be familiar with the general principles behind each experiment from a lecture course. The organization of each chapter is as follows

- *Review of principles:* Each chapter begins with a to-the-point review of basic principles (what is a nucleic acid, what does a polymerase do and how it works etc.). In the authors' experience, limiting background in a lab manual to information that is pertinent to performing the laboratory experiment is a good strategy at the undergraduate level. This approach directs students' attention to putting the theory into practice as opposed to learning the theory in lab, which can be done better in the lecture course. In other words, we find that when we teach lab with a clear focus on improving students' experimental skills we achieve a higher content retention.
- *Reagent needs:* A complete list of materials and equipment needed is listed for six student teams. Two-to-three students per team works well to maximize peer interaction while still making sure that each student has a chance to intellectually contribute to the assignments. The course was typically taught by a trained biochemist and a teaching assistant at the authors' institution, but one instructor is sufficient for a group of no more than 16 students.
- *Protocols:* A detailed protocol is given for each experiment, including recommended timeline to complete experiments within three hours, which typically is the time allocated for a biochemistry laboratory session.
- *Notes to the instructor:* This section provides information on how to (a) alter the protocol to accommodate different instrumental setups; (b) arrange the experiments to fit within the three hours lab time that is commonly used at most institutions; and (c) how to utilize the waiting time that is inevitable when performing biochemistry experiments.
- *Problem sets:* Problem sets are grouped into three categories:
  - *Pre-lab questions* are designed to focus students' attention to the most important points in the experiment. In the authors' experience, students perform much better during lab if they are asked to answer simple questions about the experiments ahead of time. Therefore, it is recommended that students complete pre-lab assignment before coming to lab.

- *Lab report checklists* contain questions that guide students through data processing and analysis.
- *Worksheets* contain problems that are designed to help students think more closely about each experiment. These questions are of increasing difficulty. The instructor may assign all the questions or pick ones that best match the skill level of the class. They work well as problem sets during lab to fill the waiting time that is notorious for biochemistry experiments.

Contents

1 Introducing the Bacterial Antibiotic Sensor Mini Project

Antibiotic resistance is an emerging problem in modern medicine: 70% of bacterial strains are resistant to at least one antibiotic, making treatment of common bacterial infections increasingly difficult. As a result, in the United States more people die from bacterial infections than from HIV infection and breast cancer combined. During the course of the mini project, you will learn about this important problem while getting hands-on experience with a wide variety of techniques currently used in academic and research settings: bioinformatics, site-directed mutagenesis, nucleic acid manipulation techniques and fluorescence binding assays. All of our experiments use a harmless model organism; therefore, there is no danger of getting infected.

1.1 What are Antibiotics?

The accidental discovery by Sir Alexander Fleming that the mold *Penicillium notatum* could destroy colonies of *Staphylococcus aureus* (STAPH) led to one of the greatest breakthroughs in the war against infectious diseases – the discovery of antibiotics. Commercially available antibiotics are modified natural compounds. Antibiotic scaffolds often come from natural antibiotics that are produced by soil bacteria (tetracycline, streptomycin) or fungus (penicillin). Bacteria make antibiotics for two distinctly different reasons. Firstly, at low concentration antibiotics act as signaling molecules that regulate the homeostasis of microbial communities, and may actually stimulate cell growth. In this context antibiotics play a role in cell-to-cell communication, also known as quorum sensing, and coordinate cell growth within the bacterial community. Second, at high concentrations antibiotics are agents of microbial warfare: they are produced by one bacterial species to kill another. To date, there are 160 classes of antibiotics known; most were discovered between 1940 and 1960.

Antibiotics exert their therapeutic function by exploiting the difference in protein synthesis between bacteria and eukaryotes. The target antibiotic of the toxin sensor we study is tetracycline (*Fig 1.1*). Tetracycline is a polycyclic aromatic compound that halts protein synthesis in bacteria while leaving protein synthesis of the host organism unaffected. As a result tetracycline antibiotics kill bacteria without harming the patient. A commonly prescribed tetracycline derivative is doxycycline, used to treat gum disease, urinary tract infection, chlamydia and gonorrhea, among other bacterial infections.

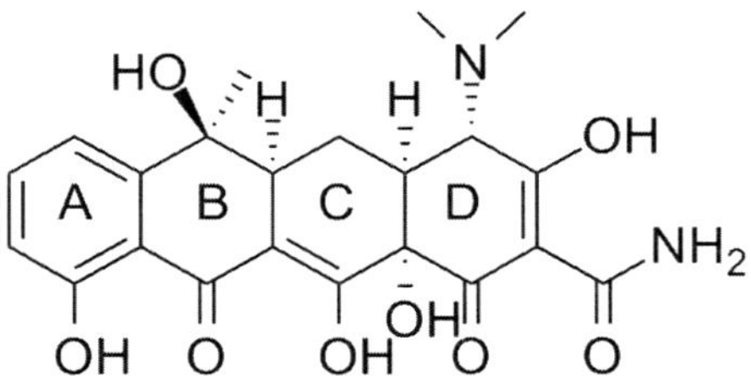

Figure 1.1: Tetracycline is a natural antibiotic produced by soil bacteria. It is a polycyclic aromatic compound.

1.2 What is Bacterial Antibiotic Resistance?

Bacterial antibiotic resistance is the ability of pathogenic bacteria to resist treatment with antimicrobial agents such as antibiotics. Antibiotic resistance is a serious threat to global health, because it jeopardizes treatment of an increasingly large number of infections caused by bacteria, fungi or a virus. According to the World Health Organization in 2012, there were about 450 000 new cases of multidrug-resistant tuberculosis. Extensively drug-resistant tuberculosis has been identified in 92 countries. Resistance to earlier generation antimalarial drugs is widespread in most malaria-endemic countries. There are high proportions of antibiotic resistance in bacteria that cause common infections (e.g. urinary tract infections, pneumonia, bloodstream infections) in all regions of the world. A high percentage of hospital-acquired infections are caused by highly resistant bacteria, such as methicillin-resistant STAPH (MRSA), or multidrug-resistant Gram-negative bacteria. Gonorrhea may soon become untreatable, because treatment failures using third-generation drugs were reported from 10 countries and no vaccines or new drugs are in development. Patients infected with a drug-resistant pathogen are at an increased risk of worse clinical outcomes (even death), and generally require more healthcare resources compared to patients infected with a non-drug resistant strain. The Center for Disease Control estimates the direct costs associated with hospital infections are as high as $45 billion dollars each year. With the emergence of bacterial strains that are resistant to multiple treatments, there is an increased urgency to understand how bacterial defense mechanisms are triggered in the presence of antibiotics.

Bacterial antibiotic resistance emerges for three reasons. (1) Antibiotics became the to-go treatment for infections and were often prescribed unnecessarily. (2) Patients sometimes do not finish their prescription, but stop taking the antibiotic once they feel better. In this case, the treatment is stopped before the infection is completely eliminated leading to propagation of resistant bacteria. (3) Bacteria have an intrinsic ability to thrive in toxic environments by becoming resistant to these toxins.

How do bacteria become resistant to antibiotics? There are four different mechanisms by which bacteria become resistant to antibiotics (*Fig. 1.2*). (A) The bacteria may degrade or modify the antibiotic. An example for this strategy is production of the enzyme β-lactamase that degrades antibiotics of the penicillin family. (B) Bacteria may undergo mutagenesis and change the target of the antibiotic. A good example are antibiotics that exert their function by directly binding to the ribosome thus halting bacterial translation. Once the relevant part of the ribosome is mutated, the antibiotic cannot bind to it and is no longer a successful therapeutic against that bacteria. (C) The antibiotic can be prevented from entering the cell. Mycobacterium is typically more resistant to treatment, because its waxy cell envelope prevents entry of antibiotics. (D) Antibiotics may be removed from the cell with the help of efflux pumps. Efflux pumps are integral membrane proteins that pump toxins out of the cell. They are either specific to one drug or pump multiple drugs out of the bacteria. The latter type is called multidrug-resistance efflux pump or MDR pump. The toxin sensor studied in the mini project regulates expression of a MDR efflux pump in *Bacillus subtilis*. Counterparts of the sensor are found in many Gram-positive bacteria.

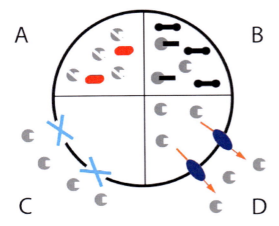

Figure 1.2: Bacteria render antibiotics ineffective using one of four strategies: they degrade the antibiotic (A); they alter the target through mutation (B); they block the entry of antibiotics (C) or pump them out of the cell with the help of efflux pumps (D).

1.3 How Do the Bacteria Detect Antibiotics In Its Environment?

Just like we do - bacteria have sensors to detect toxic compounds. Once these sensors detect an antibiotic they trigger expression of a protein that renders the antibiotic ineffective as a therapeutic using one of the four mechanisms discussed above. Toxin sensors in bacteria are usually proteins (transcription factors) that specifically bind to the toxin. The toxin-sensor protein complex then binds to the bacterial DNA or mRNA to upregulate the expression of the resistance gene that renders the antibiotic harmless to the bacteria.

In contrast, the sensor studied in the mini project is actually an RNA molecule: the ykkCD sensor RNA from *Bacillus subtilis*. The ykkCD sensor is encoded in the bacteria's DNA next to a multidrug-resistant efflux pump that is also called ykkCD. The ykkCD sensor specifically recognizes the antibiotic tetracycline. The ykkCD toxin sensor binds to the antibiotic tetracycline. Tetracycline is toxic to bacteria and is a ligand of the efflux pump. Expression of the efflux pump is regulated by the ykkCD sensor. Binding of tetracycline initiates a structural change in the mRNA. This structural change enables transcription and translation of the efflux pump gene that in turns pumps tetracycline out of the bacterial cell. As a result the antibiotic tetracycline cannot be effective as therapeutic against bacteria that have the ykkCD toxin sensor or its homolog. Thus bacteria that have the ykkCD toxin sensor are resistant to tetracycline.

1.4 How Does the ykkCD Sensor Exert Its Function?

The ykkCD RNA sensor is a riboswitch. Riboswitches are conserved elements in the mRNA that regulate gene expression by allosteric structural changes. They are located in the 5' untranslated region (5'UTR) of the gene they regulate. Most riboswitches characterized to date turn off expression of a metabolite-producing gene once sufficient amounts of metabolite is synthesized. As metabolite concentration reaches a threshold, it is able to bind to the riboswitch and initiate a structural change in the mRNA that prevents expression of the metabolite-producing gene either by halting transcription or preventing protein synthesis. The main difference between gene expression regulation by riboswitches and regulation by transcription factors is that riboswitches are part of the mRNA and are able to directly bind to their target ligand without the help of a protein cofactor. *How does the ykkCD sensor undergo structural change upon tetracycline binding and directs expression of the ykkCD efflux pump?* (A) In the absence of tetracycline the ykkCD sensor RNA folds into a structure that contains a terminator stem. This stem prevents synthesis of the ykkCD efflux pump mRNA. As a result the efflux pump is not made. (B) In the presence of tetracycline the ykkCD RNA sensor folds into a structure that does not contain the terminator stem. As a result the ykkCD efflux pump is made and able to pump tetracycline out of the cells (*Fig. 1.3*).

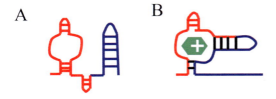

Figure 1.3: A model depicting the conformational change of the ykkCD toxin sensor upon tetracycline binding. When tetracycline levels rise to a critical threshold the ykkCD sensor binds to tetracycline and undergoes a structural change that permits production of the MDR pump.

1.5 What Do We Do During the Mini Project?

The mini project gives you a taste of hypothesis-driven real-life research and introduces you to a very important problem in biochemistry: how macromolecules recognize their specific ligand. Enzymes, transporters and receptors are all required to distinguish their specific target from other molecules. They accomplish this by forming specific noncovalent bonds (H-bonds, ion pairs and/or van der Waals interactions) with their target. The macromolecule is able to form more noncovalent interactions with their target than with other molecules therefore, the complex between the macromolecule and its target (cognate complex) is lower in energy (low dissociation constant [K_D]) than the complex between the macromolecule and other molecules (non-cognate complex). Consequently, nonspecific molecules are rejected, because they cannot form all required noncovalent interactions. This selection strategy is depicted in the schematic below (*Fig. 1.4*).

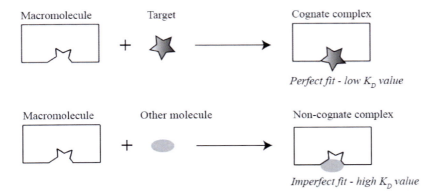

Figure 1.4: Schematics depicting specific target recognition by a macromolecule. The specific target is able to form all noncovalent interaction with the sensor leading to a low K_D value for the sensor-target complex (top). Any other molecule is rejected, because not all noncovalent interactions can be formed (bottom).

Typically only a subset of nucleotides or amino acids in the sensor is required to recognize the specific target. *How do we know which part of the macromolecule is required to recognize the target?* Usually, the part of the macromolecule that is essential for recognition is evolutionary conserved. This means, it is a common research strategy to identify elements in a macromolecule that are conserved throughout evolution, change these elements by using site-directed mutagenesis and test whether the mutated macromolecule retains its ability to recognize the target ligand by performing binding assays. If the mutated macromolecule is still able to recognize the target ligand then the regions changed are most likely were not necessary to recognize the target. In contrast, if the mutated macromolecule is no longer able to recognize its target then the region changed was probably important for recognition.

In the mini project you will set out to better understand the molecular basis of how toxin sensors recognize their target. In the process you will learn state-of-the-art techniques (listed in parenthesis in the outline below). These techniques are routinely used in biochemistry and molecular biology laboratories in industry and in academia. To determine which part of the ykkCD sensor is essential for tetracycline recognition we will follow the outline below:

(1) Identify elements within the toxin sensor that did not vary throughout evolution (sequence alignments and structure prediction).
(2) Modify these elements using site-directed mutagenesis (primer design and PCR).
(3) Synthesize the modified (mutated) sensors. This process involves several steps:
    a. Synthesis of plasmid DNA that contains the mutated sensor RNA sequence (plasmid prep)
    b. Enzymatic cut of this plasmid DNA (restriction digest)
    c. *In vitro* synthesis of sensor (RNA transcription)
    d. Purification of the mutant sensor RNA
(4) Evaluate whether that part of the toxin sensor you choose to change was essential for tetracycline recognition or not (fluorescent binding assay). If the mutated sensor still binds strongly to tetracycline (low K_D value) then the regions mutated were most likely not important for recognition. Likewise if the mutated sensor no longer binds strongly to tetracycline then the regions changed were likely to be important for recognizing tetracycline (high K_D value). This thought process is illustrated on *Fig. 1.5*:

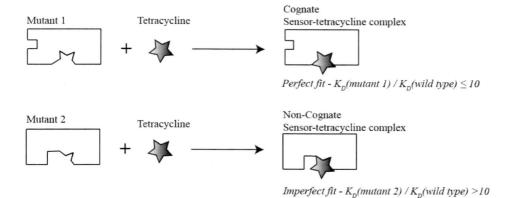

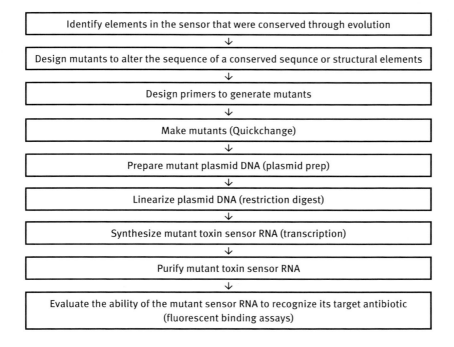

Figure 1.5: Schematics depicting how to evaluate the results of the ykkCD RNA mutagenesis study. If the nucleotides mutated were not important for tetracycline recognition then all important noncovalent interactions still form between tetracycline and the sensor and thus a low K_D value is measured (top). If the nucleotides mutated were important for recognition then not all noncovalent interactions are able to form between the mutant sensor and tetracycline. As a result a high K_D value is measured (bottom).

A flowchart of the mini project is as follows:

Identify elements in the sensor that were conserved through evolution
↓
Design mutants to alter the sequence of a conserved sequnce or structural elements
↓
Design primers to generate mutants
↓
Make mutants (Quickchange)
↓
Prepare mutant plasmid DNA (plasmid prep)
↓
Linearize plasmid DNA (restriction digest)
↓
Synthesize mutant toxin sensor RNA (transcription)
↓
Purify mutant toxin sensor RNA
↓
Evaluate the ability of the mutant sensor RNA to recognize its target antibiotic (fluorescent binding assays)

Note to the instructor
Chapter 1 is designed to serve as introduction to the antibiotic resistance mini project. This project was performed in eight laboratory sessions that lasted 3 hours each. It is advisable to give the introductory lecture before the start of the eight-week segment so that students are familiar with the project before they embark on it. The introductory lecture can be conveniently combined with check-in and/or safety training that often takes place when undergraduate labs meet the first time.

2 Identifying Conserved Elements in the Toxin Sensor and Designing Mutants to Test Whether They are Important for Function

2.1 Learning Objectives

During this lab, you will use bioinformatics to learn how to find the DNA sequence of a given macromolecule and use this sequence to uncover evolutionary sequence conservation. You will use these data to identify *conserved sequence segments* (invariable blocks) in the ykkCD sensor RNA. During the second half of the lab you will identify *conserved structural elements* within the toxin sensor. These are elements where the sequence may have been altered during evolution, but the structure was retained. You will then use this information to design a mutant to see if a conserved sequence or structure is important for toxin recognition.

2.2 Mini Project Flowchart

The bolded blocks in the flowchart below highlight the role of the current experiment in the mini project.

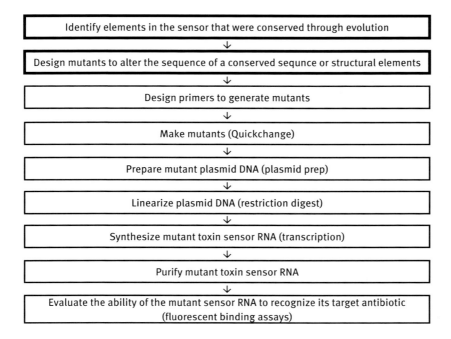

2.3 Why is Sequence Conservation Important for Macromolecule Function, and How Do We Determine This?

As you learned earlier, the goal of the mini-project is to better understand the molecular basis of how the ykkCD sensor RNA recognizes the antibiotic tetracycline. You will determine which part of the sensor is essential for tetracycline recognition. The first step toward this goal is to identify segments of the toxin sensor (ykkCD riboswitch) that did not change throughout evolution. We call these segments *invariable blocks*. These elements are the most likely to play substantial roles in tetracycline recognition. You will subject these elements to site-directed mutagenesis and evaluate how these mutations affect tetracycline recognition by the sensor. *Site-directed mutagenesis* simply means that you will alter the sequence of the sensor in the bacterial DNA. If the mutation abolishes the ability of the sensor to recognize tetracycline, then the mutated part of the sensor is essential for tetracycline recognition (*Fig. 2.1*). You will conclude this if the K_D (dissociation constant) value of a mutant sensor RNA – tetracycline complex is at least 10-fold larger (recall larger K_D value means weaker binding) than that of the wild-type sensor tetracycline complex. If the K_D value of the mutant sensor is within an order of magnitude of that of the wild-type sensor tetracycline complex you will conclude that the nucleotides altered were not essential for recognizing tetracycline.

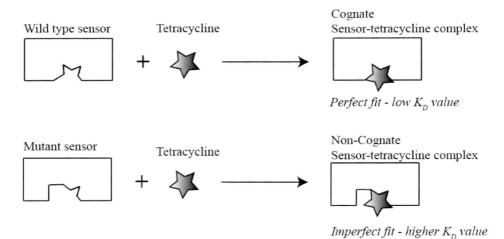

Figure 2.1: Evaluation of ykkCD sensor RNA mutants using binding affinity assays: If $\frac{K_D^{mutant}}{K_D^{wildtype}} > 10$ (higher K_D value) the nucleotide(s) mutated were important for tetracycline recognition. If $\frac{K_D^{mutant}}{K_D^{wildtype}} < 10$ (low K_D value) the nucleotides mutated were not important for tetracycline recognition.

2.4 Review of Nucleic Acid Properties

Before we identify conserved sequence elements and design mutants let us review a few things about nucleic acids. Nucleic acids are macromolecules that contain a chain of nucleotides connected by covalent bonds (phosphodiester bonds). Most nucleic acids contain all four nucleotides: adenine (A), thymine (T) or uracil (U), cytosine (C) and guanine (G). Our toxin sensor is an RNA molecule therefore it contains uracil instead of thymine, but since we perform our mutagenesis in the bacterial DNA we will see T instead of U in the sequence. The *sequence of the nucleic acid* is determined by the order in which the nucleobases follow each other in the macromolecule, and it determines the properties of the nucleic acid. For example even though the following two nucleic acids, ATCG and GTCA, contain the same bases, since the bases follow each other in different order, the properties of these two nucleic acids will be very different. *Secondary structure*, the interaction pattern between nucleobases, of a nucleic acid is easily predictable since an A always pairs with a T or a U using two hydrogen bonds and a C always pairs with a G using three hydrogen bonds. These pairs are referred to as *Watson-Crick pairs or base pairs*. Nucleic acid sequence comparisons and structure predictions are important applications of bioinformatics.

2.5 What is Bioinformatics?

During the last 15 years one of the greatest breakthroughs in biochemistry was the emergence of bioinformatics. *Bioinformatics* uses the power of computer science to analyze biology data and generate new information. This new field of life sciences has emerged, because (1) there has been an exponential increase in biological information due to genome sequencing, gene expression profiling (microarray data) and determination of macromolecular 3D structures, and (2) there has been a vast improvement in the access to computational power. The most common applications of bioinformatics in biochemistry are:

1. Performing sequence comparisons to determine macromolecule function by comparing the sequence of a new macromolecule to that of macromolecules with known function.
2. Generating sequence alignments to identify regions of a macromolecule that are conserved through evolution. These regions are likely to be important for function.
3. Predicting secondary or tertiary structure of a macromolecule by comparing its sequence to macromolecules with known structure.
4. Predicting and visualizing macromolecule-ligand interaction to determine how the macromolecule recognizes its specific target.

During this lab course we will familiarize ourselves with most of these applications.

Genome sequences are stored in the gene databank (GenBank) of the National Institute of Health (NIH). GenBank is the annotated collection of publicly available gene sequences. By the end of 2013 there was 169,331,407 sequences stored in the GenBank, which represents a 17-fold increase in data since the sequencing of the human genome in 2000 and an about 280,000-fold increase in data since the creation of GenBank in 1982. This database is available online free of charge to perform sequence comparisons, also called sequence alignments. There are two types of sequence alignments: global alignment and local alignment. *Global sequence alignment* aims to align as many characters as possible between the sequence of interest, the query, and sequences in the GenBank, the subject. The goal is to find a hit with high overall similarity. This method is slow, but is useful for example to uncover evolutionary relatedness between two species. In contrast, *local sequence alignment* focuses on stretches of high similarity, and thus, it compares discrete parts of the sequence of interest to sequences in the GenBank. *Pairwise alignment* aims to find the best way to match two sequences. *Multiple sequence alignment* compares the sequence of interest to many other sequences in the GenBank to find regions of the sequence that are conserved through evolution (*Fig. 2.2*).

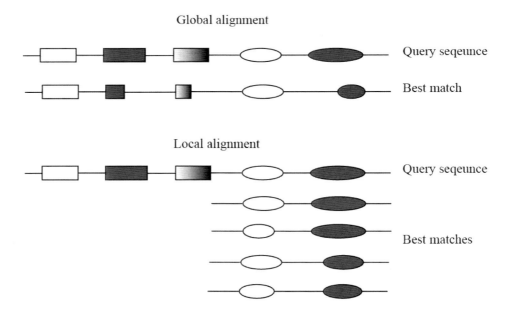

Figure 2.2: Global alignment versus local alignment: Global sequence alignment aims to find a GenBank sequence that shows significant overall similarity to the query sequence. Local sequence alignment attempts to find a GenBank sequence that shows discrete regions of significant similarity.

The two major problems encountered when performing sequence alignments are: (1) Due to the vast amount of data available in the GenBank, it takes considerable time to perform a thorough comparison, and (2) sequences can be similar at random. To overcome these problems (a) sequences are divided into short segments (words) and these segments are simultaneously compared to sequences in the GenBank and (b) alignments are scored using a scoring function. The most commonly used alignment algorithm is **B**asic **L**ocal **S**equence **A**lignment **T**ool or BLAST. *BLAST* uses a 7-15 nucleotide word size. Every time there is a nucleotide match between the query and the subject a +1 is added to the score. If there is nucleotide mismatch, a -2 is subtracted from the score. Often it is useful to introduce breaks (gap) into the alignment to generate a better overall match between two words. Introducing a gap results in a -3 penalty while extending a gap results in a -1 penalty. The alignment with the highest overall score is the best.

An example of how alignments are scored is seen below:

Query　　CCC　　　Score = 1 match + 2 mismatches = (+1) + 2 x (-2) = -3
Subject　GGCC

After shifting the query to the right:

Query　　　CCC　　Score = 2 matches + 1 mismatch = 2 x (+1) + 1 x (-2) = 0
Subject　GGCC

The second alignment has the highest score This means, it is the best.

An example of how the introduction of gaps improves alignments is seen below:

Query　　AGCAC　Score = 2 matches + 2 mismatches = 2 x (+1) + 2 x (-2) = -2
Subject　AGAC

After introducing a gap

Query　　AGCAC　Score = 4 matches + gap = 4 x (+1) + 1 x (-3) = -1
Subject　AG_AC

The alignment with the gap is a better match between the two sequences. BLAST divides the sequence of interest to words. Once the highest scoring arrangements are found the alignment is extended to find the best overall alignment. To account for similarity between sequences that takes place at random the *Expected value* or *E-value* is introduced. The *E-value* is a parameter that describes the number of random hits with a particular word size. Essentially the *E-value* represents the background noise (significance threshold) of an alignment. The closer the *E-value* is to zero the better is the alignment.

　　　Secondary structure prediction of nucleic acids is straightforward, because secondary structure formation in nucleic acids follows the simple rule of Watson-Crick base pairing: A pairs with T or U using two H-bonds whereas G pairs with C using three H-bonds, thus G-C pairs are more stable than A-T pairs. Among the potential secondary structures the one with the lowest free energy is the most likely structure. The most popular algorithm to predict RNA structure is *Mfold*, where *M* stands for multiple, meaning that it generates more than one potential structure and fold stands for structure.

2.6 Identifying Conserved Sequence Elements (Invariable Blocks)

First you will use the sequence of the sensor from the model organism *Bacillus subtilis* to find toxin sensor sequences in different organisms. Then, you will compare the sequences of the toxin sensor in these different organisms using *multiple sequence alignment* to identify blocks in the sequence that did not change throughout evolution. Once you have identified the invariable blocks of the toxin sensor, you will choose one that you subject to mutagenesis. You will delete or insert nucleotides into the invariable block of your choosing, or modify its sequence.

2.7 Identifying Conserved Structural Elements

Besides invariable blocks, conserved structural elements may also serve as hot spots for recognition. These are regions in the molecule that may have different sequence, but form the same structure. For example, both the GGGG AAAA CCCC and the AAAA GGGG TTTT nucleic acids form a hairpin even though they have very different sequence. To identify conserved structural elements in the ykkCD toxin sensor first you will predict the secondary structure of the ykkCD sensor RNA using ykkCD sequences from different organisms. Then you will select structural elements that are present in each organism based on visual inspection of the predicted structures. To test if a conserved structural element is important for toxin recognition, you will design a mutant that significantly alters a conserved element, then design another mutant that restores the original structure (compensatory mutation). If the structural element was important for toxin recognition then significantly changing the structure would destroy recognition, but restoring the original structure, albeit with altered sequence, will restore the ability of the sensor to recognize its target toxin.

PROCEDURES
Identifying invariable blocks

1. Go to http://blast.ncbi.nlm.nih.gov/Blast.cgi.
2. Click nucleotide blast.
3. Paste the sequence below into the "Enter Query Sequence" window.

T G T A A A G T T T T C T A G G G T T C C G C A T G T C A A T T G A C A T G G A C
T G G T C C G A G A G A A A A C A C A T A C G C G T A A A T A G A A G C G C G T
ATGCACACGGAGGGAAAAAAGCCCGGGAGAG

4. Choose "others" under "Choose Search Set/Database" and "Somewhat similar sequences" under "Program Selection".
5. Hit BLAST.
   You will see many sequences color-coded by the degree of similarity to the *B. subtilis* ykkCD RNA.
   Find ykkCD sequences in the organisms below and paste them into your lab notebook!
   Bacillus subtilis, Bacillus amyloliquefaciens, Bacillus licheniformis, Bacillus halodurans, Bacillus pumilus, Alkaliphilus oremlandii, Staphylococcus saprophyticus subsp, Symbiobacterium thermophilum
6. Perform multiple sequence alignment on the selected sequences by going to http://bioinfo.genotoul.fr/multalin/multalin.html.
7. Each sequence should be entered in a different row with designation preceding the sequence as follows:
   >Sequence name (for example *B.subtilis*)
   sequence (for example Tgtaaagt......)
   >Sequence name
   sequence
   etc.

8. Hit "Start MultAlin".
9. Nucleotides colored in red represent invariable groups.

Identifying conserved structural element
1. Go to http://mfold.rna.albany.edu/?q=mfold/RNA-Folding-Form.
2. Paste the RNA sequences below into the program window to predict their structure.

B subtilis sequence
UGUAAAGUUUUCUAGGGUUCCGCAUGUCAAUUGACAUGGACUGGUCCGAGAGA
AAACACAUACGCGUAAAUAGAAGCGCGUAUGCACACGGAGGGAAAAAAGCCCG
GGAGAG

Staphylococcus saprophyticus sequence
AAAACUGGCUUCUAGGGUUCCGGUCCCGCUCCUGUGGUGGGACGGCUGGUCC
GAGAGAAGCA.GCCG..GUCCGACAGCAGGGCCGGUCACACGGCGGGAGAAAA
GCCCGGGAGAG

Gloeobacter violaceus

AAUAAAGCUUUCUAGGGGUUCCGCAAGGUGAUUACUUUGGUCUG.GU
CCGAGAGAAAGCCACAUAUUUUUAUGUGACACGGAAGGAUAAAAGCCUGGGAGAU

3. Choose "Fold RNA".
4. To view predicted structures go to "View Individual Structures". Select jpeg format.

Mutant design (design three mutants)

1. Propose a mutant that changes a conserved *sequence* element of the toxin sensor (mutant 1).
2. Predict the structure of this mutant using Mfold.
3. If the mutant changed the structure of the sensor significantly (predicted structure is completely different from the original one) choose a different mutant until you find one that changed a conserved sequence without significantly changing the overall structure.
4. Design mutants to test if a conserved structural element is important for toxin recognition by following the outline below.
– Propose a mutant that significantly alters or eliminates a conserved *structural* element (mutant 2).
– Design a compensatory mutant to restore the original structure (mutant 3).

Notes to the instructor

The experiment in Chapter 2 is designed to identify sequence conservation in the *B subtilis* tetracycline sensor RNA ykkCD. The same protocol with minimal modifications could be used to identify sequence conservation in any nucleic acid (regulatory RNA, promoter or ribozyme). Two students per computer works well to maximize peer interaction while still making sure that each student has a chance to intellectually contribute to the assignments. Tablets or smart phones may also be used to complete each task. This means, this experiment may be used as an assignment in a lecture course. The websites listed above are free to use and have access to the latest entries in the nucleic acid sequence data bank, but the following websites can also be used as alternates if deemed necessary:

http://embnet.vital-it.ch/software/ClustalW.html (CLUSALW to perform sequence alignments)

http://rna.tbi.univie.ac.at/cgi-bin/RNAfold.cgi (Vienna Package to predict nucleic acid secondary structure)

BLAST Prelab

Nucleic acids review

1. Nucleic acids are polymers of _____. (/ 1 pt.)

2. Nucleic acid polymers (strands) go from ___ to 3' direction. (/ 1pt.)

3. GC pairs have _____H-bonds while AT pairs have _____H-bonds, therefore
   AT pairs are _____ (choose more or less) stable than GC pairs. (/.3 pts.)

4. Draw the structure of ATP. (/ 1pt.)

5. Indicate sugar, phosphate and the nucleobase on the nucleotide below. (/3 pts.)

6. Write the complementary sequence for the following nucleotide. (/ 1 pt.)
        5' GCCGATA 3'
        3'   5'

Questions regarding "Identifying invariable blocks in the toxin sensor"

1. Circle "invariable blocks" in the following sequence alignment. (/2pts.)

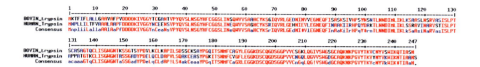

2. Circle conserved structural elements (structures that all of these RNAs have in common). (/ 2pts.)

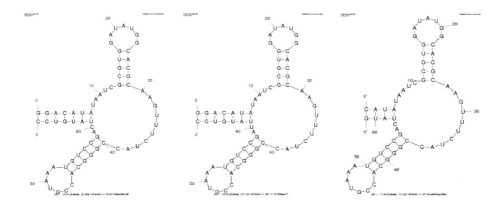

Identifying Invariable Blocks in the Toxin Sensor
Lab Report Outline and Point Distribution

1. Define the goal of the experiment (3 pts.).

2. Copy of multiple sequence alignment (3 pts.).

3. Circle each invariable block (3 pts.).

4. Secondary structure predictions for *Bacillus subtilis*, *Staphlococcus saprophyticus* and *Gloebacter violaceus* ykkCD sensor RNAs. (3 pts. each = 9 pts. total).

5. Circle the conserved structural elements for the three structural predictions performed (3 pts.).

6. Show structural prediction for mutant 1 (2 pts.).

7. Comment on why you believe that your mutation caused these specific structural changes (5 pts.).

8. Show the structural prediction for mutant 2 and mutant 3 (4 pts.).

9. Comment on why you believe that mutant 3 restored the original structure (5 pts.).

10. Based on your experience with mutants 1 and 2 explain why changing structure and sequence at the same time makes it hard to interpret the functional effect of a mutation (aka what was the reason for designing mutant 3) (7 pts.).

11. Choose a mutant (from the three designed here) that you would like to test experimentally. Provide a brief explanation for why you chose this mutation (6 pts.).

BLAst Problem Set

<u>Define or describe the following terms as used in BLAST:</u>

1. Word

2. E (Expect or Expectation value)

3. Gap

4. Max Score

<u>Find and report the following:</u>

1. The number of identities in the alignment between the query and the subject, *Cyanothece sp. PCC7425*

2. The percent of coverage between the query and the subject, *Sulfuricurvum kujiense DSM16944*

3. The number of gaps in the alignment between the query and the subject, *Clostridium kluvyeri DSM555*

4. The Expect value in the alignment between the query and the subject, *Exiguobacterium sibiricum 255-15, complete genome.*

5. The percent of maximum identity between the query and the subject, *Geobacillus sp Y412MC52*

Find and interpret the following:

1. Find the E value for the alignment between the query and the subject, *Bacillus subtilis subsp. spizizenii str. W23, complete genome.* How do you specifically interpret this number? (What does this number mean?)

2. Find the E value for the alignment between the query and the subject, *Pseudomonas stutzeri DSM 4166.* How do you specifically interpret this number? (What does this number mean?)

3. For the alignment between the query and the subject, *Paenibacillus sp. JDR-2, complete genome*, there are fourteen gaps reported yet only five gap openings are shown in the alignment. Is this a mistake? Briefly explain.

General BLAST interpretation:

1. The search using megablast gave far fewer matches than the search using blastn. Check the Algorithm Parameters for both searches. Could the difference in word length explain the difference in matches? Briefly explain.

2. The alignment with the subject, *Clostridium cellulovorans 743B complete genome*, has a smaller % identity than the subject, *Lysinibacillus sphaericus C3-41, complete genome*. Yet, these two subjects have the same maximum score. Briefly explain why this is the case.

Protein Properties Worksheet

In the past 10 years, one of the greatest improvements in biochemistry took place in the field of bioinformatics. Search engines and prediction tools enable us to easily learn a lot about a protein or RNA of interest before we actually start working on a project. The availability of these online tools makes it much easier to tailor a project and assess its feasibility. This problem set introduces you to a few easy-to-use prediction tools. You will learn how to calculate the molecular weight, isoelectric point or amino acid composition of a protein, how to perform sequence alignments to assess sequence conservation, how to predict secondary and tertiary structure.

Follow the step-by-step instructions below and answer all questions

You will use the sequence of the ykkCD multidrug-resistance efflux pump. Efflux pumps are an important part of bacterial defense against antibiotics: they pump antibiotics out of the bacterial cell thereby rendering them ineffective in treating bacterial infections. The RNA sensor (ykkCD riboswitch) you perform mutagenesis on in the biochemistry lab turns on production of the ykkCD efflux pump.

The ykkCD efflux pump is a heterodimer pump meaning that it is made out of two different proteins: ykkC and ykkD. **In the first part of the exercise you will perform a series of predictions using these sequences to get an idea about the physico-chemical and structural properties of this pump.**

The sequences of the ykkC and ykkD pumps are listed below:

>ykkC
MKWGLVVLAAVFEVVWVIGLKHADSALTWSGTAIGIIFSFYLLMKATHSLPVGTVYAVF
TGLGTAGTVLSEIVLFHEPVGWPKLLLIGVLLIGVIGLKLVTQDETEEKGGEA
>ykkD
MLHWISLLCAGCLEMAGVALMNQYAKEKSVKWVLLIIVGFAASFSLLSYAMETTPMG
TAYAVWTGIGTAGGALIGILFYKEQKDAKRIFFIALILCSAVGLKILS

1. Generate a sequence alignment between the two pump monomers (C and D) using this link: http://npsa-pbil.ibcp.fr/cgi-bin/npsa_automat.pl?page=npsa_clustalw.html. Make sure you copy/paste both sequences into the program as it is shown above (names and > included).
   a) Paste the sequence alignment!
   b) How similar are these two sequences (% conservation)?
   c) Is there any trend in the types of amino acids that appear conserved (apolar/polar/charged)?

d) Based on "c" what prediction would you make about the structural similarity of these two proteins (similar or not and why)?

2. Choose *the ykkC* or *the ykkD pump* to perform a series of predictions using the link below http://www.expasy.ch/tools/protparam.html; pI (isoelectric point), MW and amino acid composition. *You may use the ykkC or the ykkD pump sequence for this excersize!*

   a. List pI (isoelectric point) and molecular weight (MW) of this protein.
   b. What are the three most common amino acids in this protein?
   c. What can you tell about the general physical properties (polar/apolar; acidic/basic) of this protein? Would you expect this protein to be water soluble?

3. Predict the secondary structure of the ykkC or the ykkD pump by using:
   http://www.compbio.dundee.ac.uk/www-jpred/

   a. Paste the result of the secondary structure prediction into your report!
   b. What is the major secondary structural motif in this protein?
   c. Based on your secondary structure prediction which amino acids are prone to form a-helix?

4. Using the link below predict membrane topology of the pump used in "3".
   http://phobius.sbc.su.se/. Answer the questions below

   a. Paste the prediction here.

   b. Based on the secondary structure prediction in "3" and the previous topology prediction sketch how the pump is inserted into the membrane.

5. **Tertiary structure prediction:** Below is the sequence of the MexA multidrug resistance efflux pump from the pathogenic organism *Pseudomonas aeruginosa*. This organism mostly affects patients with compromised immune system. It is the main culprit behind infections caused by medical implants (like catheter).

Perform tertiary structure prediction on this protein using the link and the sequence below!
http://swissmodel.expasy.org/workspace/index.php?func=modelling_simple1

Warning! Performing this modeling will take a while (10-20min). Do not use the back button on your browser; just wait patiently until it is done.

>MexA
MQRTPAMRVLVPALLVAISALSGCGKSEAPPPAQTPEVGIVTLEAQTVTLNTEL
PGRTNAFRIAEVRPQVNGIILKRLFKEGSDVKAGQQLYQIDPATYEADYQS
AQANLASTQEQAQRYKLLVADQAVSKQQYADANAAYLQSKAAVEQARINLRY
TKVLSPISGRIGRSAVTEGALVTNGQANAMATVQQLDPIYVDVTQPSTAL
LRLRRELASGQLERAGDNAAKVSLKLEDGSQYPLEGRLEFSEVSVDEGTGSVTIRAV
FPNPNNELLPGMFVHAQLQEGVKQKAILAPQQGVTRDLKGQATALVVNAQNKVEL
RVIKADRVIGDKWLVTEGLNAGDKIITEGLQFVQPGVEVKTVPAKNVASAQKADAA
PAKTDSKG

a. Paste the picture of the resulting structural model into your report!

b. What can you tell about the distribution of polar and apolar amino acids in a membrane protein as compared to a water soluble protein?

3 Designing Primers for Site-Directed Mutagenesis

3.1 Learning Objectives

During the next two labs you will learn the basics of site-directed mutagenesis: you will design primers for the mutants you designed earlier and perform PCR amplification to make that mutant. In this handout you will review the basics of primer design while in the next handout you will learn about PCR amplification in practice.

3.2 Mini Project Flowchart

The bolded box in the flowchart below highlights the role of the current experiment in the mini project.

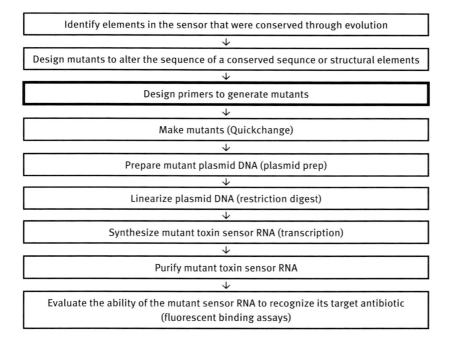

Identify elements in the sensor that were conserved through evolution
↓
Design mutants to alter the sequence of a conserved sequnce or structural elements
↓
Design primers to generate mutants
↓
Make mutants (Quickchange)
↓
Prepare mutant plasmid DNA (plasmid prep)
↓
Linearize plasmid DNA (restriction digest)
↓
Synthesize mutant toxin sensor RNA (transcription)
↓
Purify mutant toxin sensor RNA
↓
Evaluate the ability of the mutant sensor RNA to recognize its target antibiotic (fluorescent binding assays)

3.3 What is PCR? What are polymerases?

Before we begin we need to review a few definitions commonly used when we talk about site-directed mutagenesis. *Site-directed mutagenesis* means that we change, insert or delete a few nucleotides within the amino acid or nucleotide sequence. In other words we change relatively few, 4-5, nucleotides or amino acids in a macromolecule. Site-directed mutagenesis became significantly easier with the emergence of PCR amplification. *PCR amplification* means that we synthesize (make) many copies of our DNA of interest (the coding region for a protein or nucleic acid) with the help of a polymerase and a programmable machine, called the PCR machine. *Polymerases* are enzymes that synthesize nucleic acids using a nucleic acid template. For example a DNA polymerase is an enzyme that makes DNA using a DNA template. The sequence of the newly synthesized DNA will be complementary to that of the template. If the template sequence is AGGC the newly synthesized DNA will be TCCG. DNA polymerases are unable to initiate DNA synthesis on their own; they need a short nucleic acid, the primer. *The primer* is a short DNA or RNA sequence that is complementary to the template and is used to initiate DNA synthesis. The *PCR machine* can precisely cycle through temperature changes to accommodate the needs of DNA synthesis. For example the PCR machine can change the temperature from 95 °C to 68 °C precisely within a few seconds. You will learn more about the temperature changes necessary to accommodate PCR amplification and the mechanism of polymerization during the next lecture.

Following PCR amplification, the amplified DNA is digested using restriction endonucleases and ligated into a cloning vector. *Restriction endonucleases* are enzymes that cut DNA at a given sequence. For instance the restriction endonuclease *EcoRI* cuts the DNA strand every time the GAATCC sequence appears in the genome. *Ligation* means that we connect two separate nucleic acids with a covalent bond; we simply paste them together. *Cloning vectors* or *plasmids* are circular DNAs that can be replicated by the bacterial or eukaryotic host independent of replicating their own genome. This means, they allow scientists to use a bacteria or eukaryotic cell to make large amounts of the DNA that code for the protein or nucleic acid of interest. In addition, cloning vectors have features that allow easy insertion and removal of the desired DNA sequence. Bacterial cloning vectors also have a selective marker (antibiotic resistance gene). Using selective medium this marker only allow propagation of host cells that contain the cloning vector.

3.4 PCR Amplification of a Desired DNA Segment Of The Genome (Conventional Cloning)

When the project starts the first thing to do is to amplify the DNA of interest from the genome. In this section you will learn how to do that. Afterwards, you will learn how to perform site-directed mutagenesis using the Quickchange kit. Imagine you want to amplify the DNA segment below. You will need two *primers*: one is complementary to the beginning while the other is complementary to the end of the sequence. The primer that is complementary to the beginning of the double-stranded DNA (dsDNA) sequence is the *top primer* whereas the primer that is complementary to the end of the sequence is the *bottom primer*. Notice that the *top primer* anneals against the bottom DNA strand and the *bottom primer* anneals against the top DNA strand.

During the first cycle of PCR amplification you do NOT get the desired DNA segment. Instead, you get two DNAs: one of them starts at the beginning of our desired sequence, and the other ends with the desired DNA sequence. Both sequences extend beyond the DNA of interest (*Fig. 3.1*).

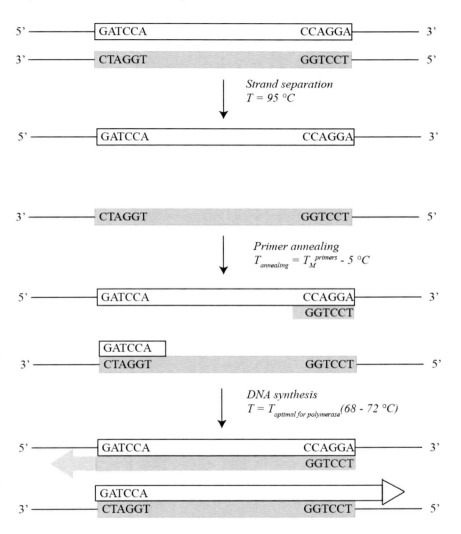

Figure 3.1: Schematics of the first PCR amplification cycle. Note that the desired DNA sequence is not generated yet.

During the second cycle of PCR amplification you finally get the product you want. The primers are more likely to anneal to the amplified DNAs than to the original template, because you have more of the amplified DNA than of the template. As a result the desired product is synthesized.

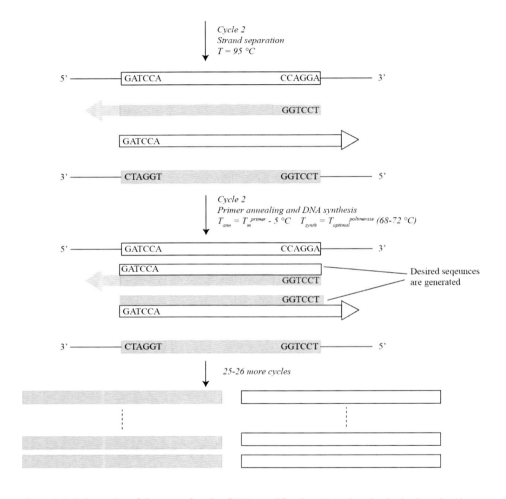

Figure 3.2: Schematics of the second cycle of PCR amplification. Note that the desired product is generated.

Notice that with each cycle the number of DNA sequences double. Thus after n cycles you have 2^n of the desired DNA sequence. Since PCR leads to significant amplification of the desired DNA, it is often called a chain reaction.

Once the DNA of interest is amplified many times, the next step is to place the desired DNA into a cloning vector. This procedure is called *cutting and pasting* and includes several steps. First, the cloning vector and the amplified DNA are digested with a pair of restriction enzymes. Second, the cloning vector is purified using an agarose gel. Third, the cloning vector is treated with the enzyme *phosphatase* to prevent it from religating without the amplified DNA. Fourth, the cloning vector and the amplified DNA is ligated together using the enzyme *DNA ligase* and transformed into cells. Not surprisingly, these steps lead to significant loss of reagents and time (*Fig. 3.3*).

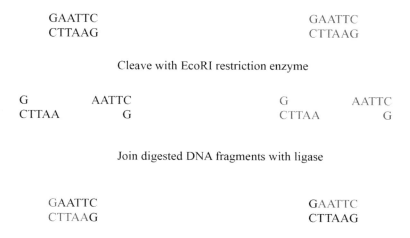

Figure 3.3: **Schematics of cutting and pasting:** During conventional cloning (cutting and pasting) both the plasmid vector and the amplified DNA has to be cleaved with the same pair of restriction enzymes and ligated together.

3.5 Quickchange Site-Directed Mutagenesis

In this section you will learn about Quickchange site-directed mutagenesis and how it differs from conventional PCR mutagenesis. Foremost, Quickchange site-directed mutagenesis does NOT require digestion with a conventional restriction endonuclease or ligation thereby reducing the time required for mutagenesis from a week to a few days. Quickchange has several restrictions. (1) Only a few nucleotides can be modified at a time. This indicates, it cannot be used to amplify a DNA sequence from the genome. (2) Quickchange provides less significant amplification of the target DNA sequence than conventional PCR. Therefore, extra care should to be taken to ensure that significant amount of mutated DNA is produced.

Let us walk through the steps of Quickchange mutagenesis (Fig 3.4)

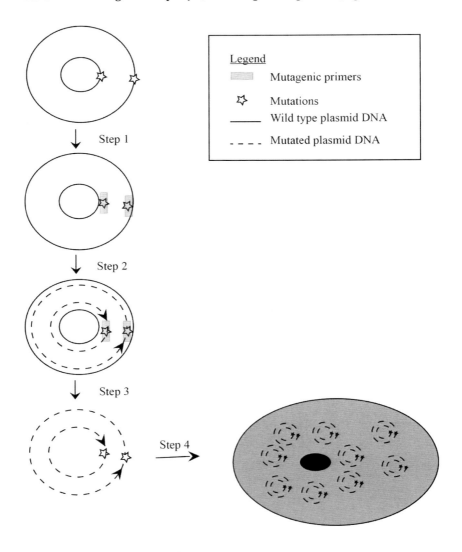

Legend
Mutagenic primers
Mutations
Wild type plasmid DNA
Mutated plasmid DNA

Figure 3.4: Schematics of Quickchange mutagenesis.

Step 1: The primers in Quickchange land at the same spot in the cloning vector. One binds to the top the other binds to the bottom strand of the double-stranded DNA. Since the polymerase replicates the entire plasmid starting from the site of mutation the target DNA sequence has to be already inserted into a cloning vector (requires circular DNA). This method cannot be used to change DNA sequence on the chromosome.

Step 2: PCR amplification makes many copies of both the top and the bottom DNA strand of the cloning vector containing the mutated DNA. Quickchange mutagenesis therefore synthesizes many copies of the *entire plasmid* not only the DNA of interest. Note that the synthesized DNA is nicked (not a full circle), this means it cannot serve as template for further PCR cycles resulting in a less significant amplification of the target sequence. Therefore, removal of the template DNA is necessary (step 3) to ensure that significant numbers of cells that harbor the mutated DNA are produced in Step 4.

Step 3: After PCR amplification the reaction mixture is treated with a unique restriction endonuclease DpnI. DpnI digests the template plasmid (the one that does not contain mutations) leaving only cloning vectors containing your mutants. The DpnI enzyme achieves this task by digesting any nucleic acid with methylated adenosine base. Nucleic acids generated with PCR do not have methylated bases; therefore they are left intact by DpnI. No purification, phosphatase treatment or ligation is necessary after DpnI treatment, thereby reducing the time and reagent needed for mutagenesis.

Step 4: The mutated cloning vector is placed into *E. coli* bacteria for further studies (transformation).

An outline of each step is shown *Fig 3.4*.

A comparison of Quickchange and conventional PCR is shown in the table below:

	Conventional	Quickchange
Primers	Complementary to beginning and end of desired sequence	Primers land at the same spot on the cloning vector
Amplification	Amplifies DNA between primers	Amplifies entire plasmid
Purpose	Site-directed mutagenesis, amplification of desired DNA from genome	Site-directed mutagenesis: only changes a few nucleotides
Time	4-5 days	1-2 days
Caveat	Slow, primer design is more complex	Only the template DNA serves as template thus it is required at a higher concentration and need to be removed prior to inserting the DNA into cells

PROCEDURES
Mutagenic primer design is illustrated below.

Primer design example:

1. Suppose that you want to mutate the highlighted G to a C:
   Tgtaaagttttctagggttccgcat**G**tcaattgacatggactgtccgagagaaaacacatacgcgtaa
   atagaagcgcgtatgcacacggagggaaaaaagcccgggagag.

2. Both primers must contain the desired mutation. The top primer anneals to the bottom DNA strand of the double stranded cloning vector. Therefore the top primer sequence will be the same, as the original sequence except it will have a C instead of a G at the appropriate spot.

3. Primers should be between 25-45 nucleotides in length with a melting temperature of T_m=78 °C. Melting temperature should be calculated using the equation below where N is the length of the primer and values of GC content should be rounded to whole numbers.

   T_m = 81.5 + 0.41*(%GC) – 675/N - % of mismatch when bases are changed

   T_m = 81.5 + 0.41*(%GC) – 675/N when bases are inserted or deleted

4. The desired mutation should be in the middle of the primer sequence with 10-15 nucleotides flanking the mutation.

5. Primers should have a GC content of at least 40%.

Mutagenic primer sequences that fulfill the requirements above for the sample sequence are ctagggttccgcatCtcaattgacatggac (top) and gtccatgtcaattgaGatgcggaaccctag (bottom). Primer sequences are always written in the 5' to 3' direction this means the top and bottom primers are reverse complements of each other. In other words that they have complementary sequences and inverse chain direction to accommodate Watson-Crick pairings, but the sequence is written in the 5' to 3' direction.

Notes to the instructor
The experiment in Chapter 3 designs primers to alter the sequence of the *Bacillus subtilis* tetracycline sensor RNA ykkCD. The same protocol with minimal modifications could be used to perform site-directed mutagenesis on any nucleic acid. Two students per computer work well to maximize peer interaction while still making sure that each student has a chance to intellectually contribute to the assignments. Tablets or smart phones may also be used to complete each task thus this experiment may be used as

an assignment in a lecture course. Students should be warned that primer design is an iterative process therefore several sequences have to be tried before a primer with the required GC content and T_M is found. Excel may be used to calculate T_M values using the equation provided. Usage of primer design programs is not recommended, because they do all the work for the students and eliminate all the educational value of this assignment.

Prelab Questions for Primer Design Lab

Define the following terms.

1. Cloning vector or plasmid.

/ 2 pts

2. DNA polymerase.

/ 2 pts

3. PCR amplification.

/ 2 pts

4. What are the pros and cons of Quickchange site-directed mutagenesis?

/ 2 pts

5. How do you calculate primer melting temperature for Quickchange mutagenesis? Outline the equation and define each term.

/ 2 pts

Introduction to Primer Design
Lab Report Outline and Point Distribution

1. Several sentences defining the goal/purpose of this experiment (3 pts.).

2. Brief description of the Quickchange mutagenesis procedure. Highlight the advantages of Quickchange over "traditional" mutagenesis (6 pts.).

3. Give your mutated primer sequence (both top and bottom) (4 pts.).

4. Report both the percentage of GC content and the T_M value (4 pts.).

5. Explain how GC content relates to the T_M of the primer. How does the T_M relate to the success of your cloning experiment (8 pts.)?

6. Explain why you chose this specific sequence for your primer. (What were you aiming for when you optimized your primer sequences?) (2 pts.).

7. How optimal is your primer? Briefly explain (3 pts.).

8. BLAST worksheet (30 pts.).

4 Performing Site-Directed Mutagenesis

4.1 Learning Objective

In this lab you will perform site-directed mutagenesis using the QuickChange mutagenesis kit (Stratagene). You will learn how polymerases work and how to amplify DNA using polymerase chain reaction (PCR).

4.2 Mini Project Flowchart

The bolded block in the flowchart below highlights the role of the current experiment in the mini project.

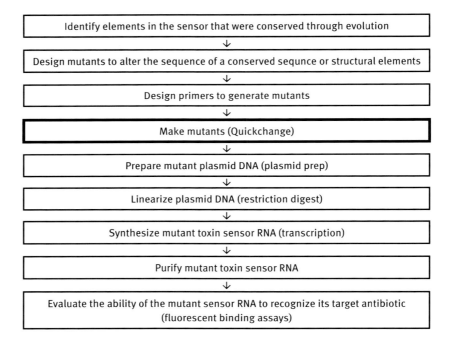

Identify elements in the sensor that were conserved through evolution
↓
Design mutants to alter the sequence of a conserved sequnce or structural elements
↓
Design primers to generate mutants
↓
Make mutants (Quickchange)
↓
Prepare mutant plasmid DNA (plasmid prep)
↓
Linearize plasmid DNA (restriction digest)
↓
Synthesize mutant toxin sensor RNA (transcription)
↓
Purify mutant toxin sensor RNA
↓
Evaluate the ability of the mutant sensor RNA to recognize its target antibiotic (fluorescent binding assays)

4.3 Review of Nucleic Acid Structure

Before you learn how polymerases work and the requirements of a successful PCR amplification you must review a few things about nucleic acid structure. *Nucleic acids* are polymers of nucleotides. The nucleotides are held together by phosphodiester bonds in the nucleic acid. Figure 4.1 shows the structure of a nucleotide. *Nucleotides* are made of a sugar moiety: ribose (RNA) or deoxyribose (DNA); a heterocyclic aromatic moiety: the nucleobase (A, C, G, T or U) and a phosphate group. When we refer to functional groups within the sugar moiety we use the apostrophe symbol; for example the second hydroxyl group on the ribose in the nucleotide is referred to as the 2'-OH group.

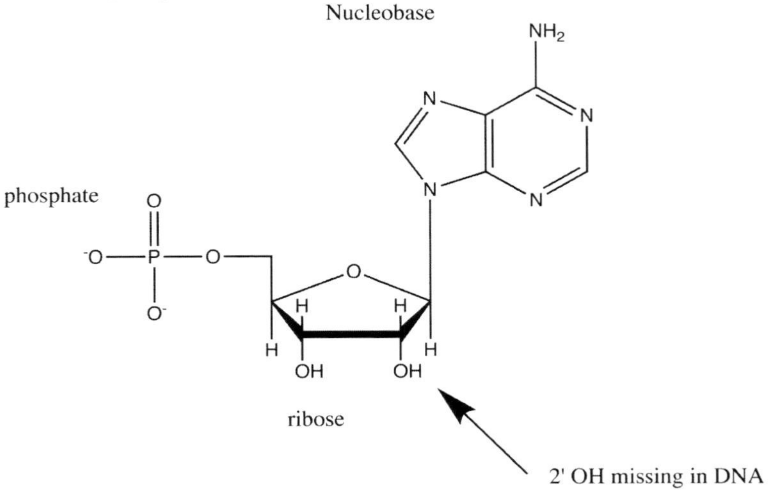

Figure 4.1: Structure of a nucleotide: Nucleotides are made out of a sugar, a nucleobase and 1-3 phosphate groups.

As we mentioned earlier, nucleotides are held together by the phosphodiester bond in the nucleic acid chain. The *phosphodiester bond* is made out of two phosphate ester bonds: each is formed between a OH group of the ribose or deoxyribose and a OH group of the phosphate. Since phosphoric acid is a moderately strong acid, the phosphodiester bond deprotonates under physiological conditions giving nucleic acids a negative charge (*Fig. 4.2*).

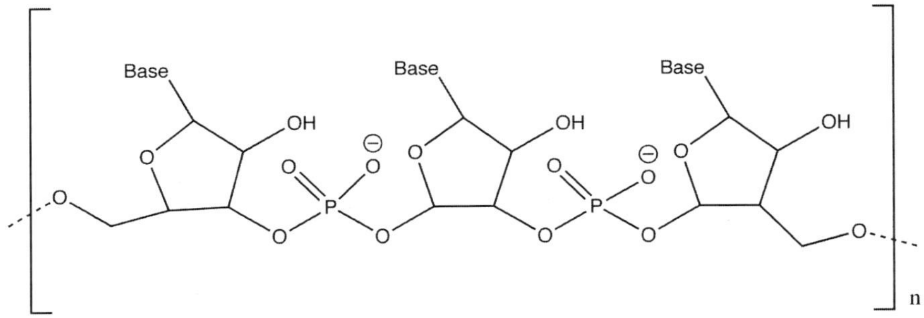

Figure 4.2: Polynucleotide structure: Nuclotides in a nucleic acid are held together by the phosphodiester bond; the chain goes in the 5' to 3' direction.

Nucleic acid chains have *polarity* just like protein chains. Nucleic acid polymers start with the 5' phosphate of the first nucleotide and end with the 3'-OH group of the last nucleotide in the chain. Therefore the chain of the nucleic acid has a precise direction; it goes from 5' to 3' direction end just like protein chains go from the N-terminus to the C-terminus. In a double-stranded nucleic acid like DNA (our genetic material), one DNA polymer goes 5' to 3' and is called the *top strand* and the other goes from 3' to 5' direction is called the *bottom strand*. These two DNA strands are complementary to each other: adenine (A) against thymine (T) and cytosine (C) against guanine (G) to ensure proper Watson-Crick base pairing between the strands. In other words the two DNA strands in this double-stranded DNA (dsDNA) are *complement* of each other. Complement means that the two strands are complementary to each other and the chain direction is opposite: the top strand goes 5' to 3' whereas the bottom strand goes 3' to 5'.

    5' AGGCCATTGGA 3'
    3' TCCGGTAACCT 5'

4.4 How do Polymerases Work?

Polymerases synthesize nucleic acids using a nucleic acid template. The sequence of the newly synthesized nucleic acid will be *complement* of the template sequence. During nucleic acid synthesis, the 3'-OH group of the growing nucleotide chain acts as a nucleophile to attack the phosphorous of the incoming nucleotide, a *pyrophosphate* (PP_i) leaves and the phosphodiester bond forms. This means that the newly synthesized DNA chain grows in the 5' to 3' direction (*Fig. 4.3*).

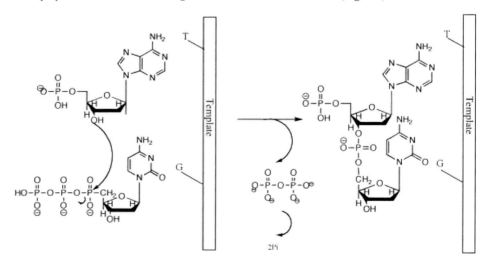

Figure 4.3: Mechanism of polymerization: The 3'-OH group of the growing nucleic acid chain acts as a nucleophile and attacks the alpha phosphorous of the incoming nucleotide triphosphate to

catalyze formation of the phosphodiester bond.

To properly position the nucleophile for attack divalent metal ions (usually Mg^{2+}) are necessary for successful DNA synthesis.

Polymerase mechanism in a nutshell:

– Polymerases synthesize DNA using a template that is the complement of the newly synthesized DNA.
– To synthesize DNA, polymerases require an OH group to act as a nucleophile. This OH comes from the growing nucleic acid chain. Recognize that the reaction is a nucleophilic substitution.
– The leaving group is pyrophosphate (PP_i), which is a high-energy molecule that splits into two inorganic phosphates. This reaction is catalyzed by the enzyme inorganic phosphatase *in vivo*.
– Polymerization is energetically favorable, because two high energy anhydride bonds are broken (one in the incoming nucleotide triphosphate and the other in pyrophosphate) and one stable bond forms (ester bond connecting the nucleotides).
– To initiate DNA synthesis, DNA polymerases require a *primer* to provide the required OH group as nucleophile.
– Polymerases travel from 3' to 5' direction on the bottom strand of the dsDNA template while they synthesize the growing chain in the 5' to 3' direction.

Figure 4.4: The newly synthesized nucleic acid has complementary sequence and opposite chain direction than the template

4.5 Polymerase Chain Reaction (PCR) in Practice

To synthesize DNA in the lab we need to perform PCR amplification of the DNA of interest using a plasmid vector or genomic DNA as template. For successful PCR amplification we have to cycle through three steps 25-30 times. The steps of PCR amplification are as follows:

1. Separation of the dsDNA template or strand separation
2. Annealing of the primers to the template DNA (they form Watson-Crick base pairs) to initiate DNA synthesis
3. DNA synthesis catalyzed by a polymerase

Each of these steps requires a different temperature. Strand separation is usually accomplished by heating the dsDNA to 95 ºC; primers form Watson-Crick base pairs with the DNA in around 55 ºC (usually 5 ºC below the melting temperature of the primer) and polymerases used in PCR work best at 68-72 ºC.

A graphic representation of a PCR cycle is seen on Fig. 4.5.

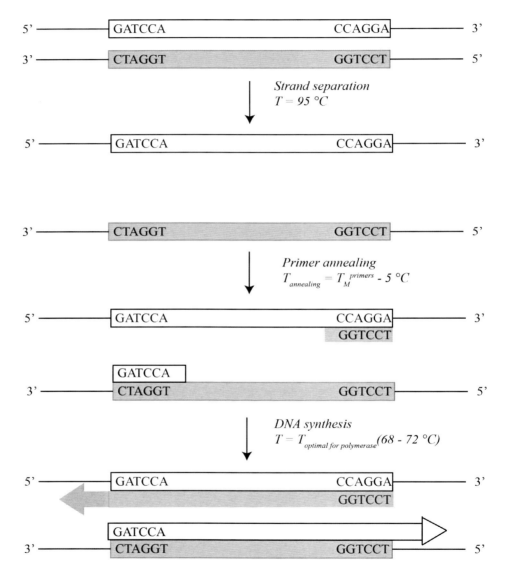

Figure 4.5: Schematic of a PCR cycle. Note that each step takes place at a different temperature so PCR machines have to be able to change temperature quickly and precisely.

4.6 Why Did PCR Only Become Widely Available in the 1980s?

PCR amplification required two scientific breakthroughs. (1) Scientist had to invent a programmable machine that can cycle through the temperatures of the PCR cycle multiple times. The PCR machine can precisely and quickly change temperature; it can go from 95 ºC to 68 ºC within a few seconds. (2) Scientists had to find a DNA polymerase that can survive the first step of the PCR cycle: incubation at 95 ºC. This step is required to separate the strands of the dsDNA. Since polymerases are proteins, most bacterial or eukaryotic polymerases would not survive this incubation (think what happens when you boil an egg). Scientists turned to organisms that live under extreme conditions: in high temperature volcanic vents. Most of these organisms belong to the third kingdom of life – *archaea*. Polymerases from these organisms evolved to be active at 60-80 ºC, therefore incubation at 95 ºC for a few minutes does not harm them. *Taq* polymerase has a half-life of 2 hrs. at 95 °C while *Pfu* Turbo polymerase has a half-life of 19 hrs. at 95 °C. In addition, polymerase used for PCR has to be of high fidelity – this means that the polymerase very rarely incorporates a wrong nucleotide into the synthesized DNA (has low error rate). *Taq* polymerase has a 16% error rate in a 1 kilo base DNA sequence while *Pfu Turbo* polymerase has a 2.6% error rate on the same sequence. In our PCR amplification you use a DNA polymerase from the Archaea *Pyrococcus furiosus* (*Pfu* Turbo).

4.7 Applications of PCR

PCR is used to amplify DNA that is present only in a few copy numbers with several orders of magnitude. As a result PCR amplification generates millions of copies of the desired DNA sequence. PCR revolutionized molecular biology, and has a wealth of applications beyond basic science. A few of those applications are highlighted below.

1. Specific DNA isolation: Using primers designed for the gene of interest a given DNA sequence can be amplified. This DNA can then be inserted into a cloning vector for further study or its sequence can be determined.
2. DNA quantification (quantitative real-time PCR or qRT-PCR). This method estimates the amount of a given nucleic acid sequence present in the sample. Since theoretically each PCR cycle doubles the amount of the given nucleic acid the fewer amplification cycles it takes to generate a detectable amount of the DNA the higher its concentration was in the original sample. This means levels of gene expression can be determined. The method requires simultaneous amplification and detection (real-time). Primers amplify the nucleic acid of interest in the presence of a fluorescent dye that is specific to DNA. The number of cycles it takes to get detectable fluorescence intensities is quantified (C_T number). This number is inversely proportional to the concentration of the target nucleic acid in the sample.

3. <u>Gene mapping.</u> Using specific primers million –fold amplification of a given part of the genome can be achieved. That way, even if only a few copy numbers of the given DNA is available, using PCR sufficient amount of DNA can be generated for scientific studies. This trait of PCR resulted in a wealth of applications beyond basic science. *DNA profiling* (genetic fingerprinting) is a technique used by forensic scientists to identify an individual involved in a crime or determine parent-child relationship between individuals (paternity testing). Even though 99.9% of the human genome sequence is shared, there are sufficient differences between each human to allow identification. The most commonly used testing procedures today focus on short tandem repeats (STR). These regions of the human genome are very variable in sequence, this means it is extremely unlikely that individuals have similar sequences unless they are closely related. Only monozygotic twins, "identical twins", have the same short tandem repeat sequences.
4. <u>PCR in diagnostics.</u> Using specific primers sensitive to a given pathogen (bacteria or virus) the source of infection can be identified using PCR much quicker than by culturing samples. Likewise, mapping of specific parts of genome can reveal if an individual is prone to breast cancer or other diseases that are much more treatable if detected at an early stage.

PROCEDURES
Reagents and equipment needs are calculated per six student teams. There is ~20% excess included.

Equipment/glassware needed:
   1. PCR machine
   2. Three sets of micropipettes 20-100 µl, 2-20 µl and 1-10 µl
   3. 6 PCR tubes
   4. 6 centrifuge tubes

Reagents needed
   1. Pfu Turbo polymerase master mix (Stratagene); 130 µl total
   2. DpnI restriction enzyme (6 µl total)
   3. Primers to generate each mutant
   4. Plasmid DNA containing the ykkCD toxin sensor

To set up PCR reaction, carefully mix the following reagents, place tubes into the PCR machine and start protocol
   1. 20 µl 2X Pfu Turbo Mastermix
   2. 100 ng plasmid DNA (volume depends on DNA concentration)
   3. 1 µl 1 µM top primer
   4. 1 µl 1 µM bottom primer
   5. Water to 40 µl

The PCR machine is programmed as follows:

Segment	Cycle	Temperature	Time	Purpose
1	1	95 °C	30 sec	Initial strand separation
2	18	95 °C	30 sec	Strand separation
		55 °C	1 min	Primer annealing
		68 °C	3 min	Polymerization
3		4 °C	infinite	

Remove template DNA (does not contain the mutation)
1. Remove reactions from the PCR machine and briefly centrifuge them to ensure that all the reaction mixture is at the bottom of the tube.
2. Add 1 μl DpnI restriction enzyme to each tube.
3. Incubate reactions in a 37 °C water bath for 1 hr.
4. Reactions may be stored at -20 °C until needed.

Notes to the instructor
The experiment in Chapter 4 is designed to perform site-directed mutagenesis on the *Bacillus subtilis* tetracycline sensor RNA ykkCD. The protocol with minimal modifications could be used to do site-directed mutagenesis on any nucleic acid. Usage of a high-fidelity DNA polymerase, such as *Pfu Turbo*, is essential for the success of the experiment, but alternative vendor or packaging may be used. Primers were ordered from Integrated DNA Technologies and reconstituted in 1 x TE (Tris-EDTA) at 1 μM concentration. Plasmid DNA containing the ykkCD sensor RNA is available upon request from the authors. Since this PCR reaction takes about two hours it is convenient to start the laboratory session by setting up the PCR reactions and giving pre-laboratory lecture while the reactions are running. This laboratory perfectly accommodates a long exam due to the minimal wet lab work and long reaction time. Depending on the time allocated for the lab session, PCR reactions may need to be removed by the teaching assistant or the instructor.

Prelab Questions for Site-Directed Mutagenesis

Define the following terms.

1. Nucleotide

| / 1 pts |

2. Nucleic acid

| / 1 pts |

3. Reverse-complement

| / 2 pts |

4. Below is the bottom strand of a dsDNA. What is the sequence of top strand? Mark the 3' and the 5' end of the top strand sequence.

3' AAGTTCAAGGC 5'

| / 2 pts |

5. Calculate how you mix your PCR reaction if the concentration of your plasmid DNA is 275 ng/µl (use the Protocol section of your handout). Show your work.

| / 4 pts |

Site-directed Mutagenesis
Lab Report Outline and Point Distribution

1. Several sentences defining the goal/purpose of this experiment (2 pts.).

2. Briefly describe the DNA polymerase reaction. Why is it thermodynamically favorable (about 5 sentences; 4 pts.)?

3. Briefly describe Quickchange site-directed mutagenesis (5-10 sentences; 5 pts.).

4. Outline the first cycle of a PCR reaction. Make sure you indicate the temperature that each cycle takes place. Briefly explain why you get significant amplification with PCR (4 pts.)?

5. Outline the first cycle of Quickchange PCR. Explain briefly why you get less amplification with Quickchange then with regular PCR (5 pts.).

6. PCR worksheet (30 pts.).

PCR Worksheet

DNA Structure:

1. (3 pts.) Draw the structure of pdApdCpdT. Label the 5´ and 3´ ends and circle each of the phosphodiester bonds in this small nucleic acid.

2. (3 pts.) Show the H-bonded pairing between adenine and its complementary base.

3. (3 pts.) A small double stranded DNA molecule is studied. Given the sequence of one DNA strand:

> GGCTACATTCGGAA

write the sequence for the other DNA strand. (Remember, sequences are written from 5' to 3' direction).

Standard PCR:

4. (3 pts.) Why is PCR described as a "chain" reaction?

5. (3 pts.) Explain why two cycles of PCR are necessary before the desired DNA product is made.

6. (3 pts.) In PCR, a set amount of DNA polymerase is added at the beginning and must remain active through all subsequent cycles. This requirement severely limits which DNA polymerases may be used. Why is this so?

QuickChange DNA Synthesis:
7. (3 pts.) Site-directed mutagenesis allows scientists to specify exactly where DNA base changes (mutagenesis) will occur. How does the QuickChange process allow us to achieve "site-directed mutagenesis?"

8. (3 pts.) Standard PCR allows for "amplification" of the DNA product. Does QuickChange allow for the same type of "amplification?" Explain.

9. (3 pts.) Standard PCR requires that a primer bind at the beginning of the DNA template sequence. In contrast, a QuickChange primer does not have to bind at the beginning of the template but can bind anywhere along its sequence. How do you explain this difference?

10. (3 pts.) Standard PCR can be used to make many copies of a newly isolated gene. In contrast, QuickChange is only useful for making base changes in a well-studied DNA sequence. What is different about the QuickChange process that makes such a difference in how it is used?

5 Purifying Mutant Toxin Sensor DNA from Bacterial Cells and Evaluating its Quality Using Agarose Gel Electrophoresis and UV Spectroscopy

5.1 Learning Objective

Last week you generated the plasmid DNA containing the mutant toxin sensor using QuickChange PCR amplification. This week you will learn how to extract and purify this plasmid DNA from *E coli* cells, and how to check the quality of the plasmid DNA using agarose gel electrophoresis and UV spectroscopy.

5.2 Mini Project Flowchart

The bolded block in the flowchart below highlights the role of the current experiment in the mini project.

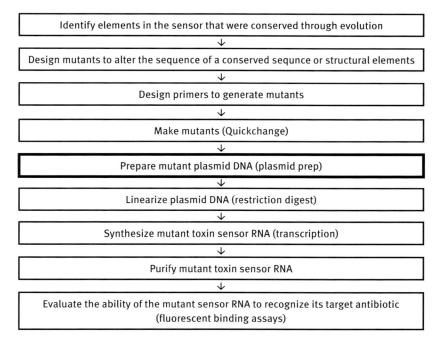

5.3 Purification of Plasmid DNA from Bacterial Cell (Plasmid Prep)

Last week, you used PCR amplification to synthesize plasmid DNA containing the mutant toxin sensor. PCR amplification is an excellent tool to change the sequence of the DNA, but it does not generate sufficient amount of plasmid DNA. PCR amplification typically yields a few hundreds of nanograms of DNA. While cloning, transfection or RNA synthesis generally requires DNA at the microgram or even milligram level. Also, plasmid DNA generated by PCR amplification is *circular* while plasmid DNA synthesized by bacteria is mostly *supercoiled*. Supercoiled DNA has the same molecular weight as its circular counterpart, but it is more compact. DNA in the supercoiled conformation is smaller, and it is better protected against harm caused by restriction endonucleases or environmental hazards (*Fig 5.1*).

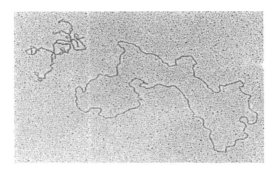

Figure 5.1: Supercoiled DNA (left) is more compact than circular DNA (right) of the same molecular weight (Courtesy of Dr. Jack Griffith).

For these reasons, PCR amplification is usually not used to make large quantities of DNA. To synthesize DNA for most applications in chemistry and biology, bacterial cells have to be used. During the *plasmid prep*, we use the DNA synthesis regime of bacterial cells, typically *E. coli*, to prepare the plasmid DNA for us. In other words, we hijack *E. coli* to synthesize our DNA.

The steps of plasmid vector synthesis using bacterial cells are as follows:
1. Transform plasmid DNA into the bacterial cell.
2. Grow bacteria containing the desired plasmid (5 to 500 ml).
3. Purify plasmid DNA from bacterial cells.

5.4 Transformation

In order to use bacterial cells to synthesize plasmid DNA, the plasmid DNA has to be placed inside the bacterial cell; meaning we have to force bacteria to take up the plasmid DNA. This is problematic, because the bacterial cell membrane is apolar and represents an impermeable barrier to polar molecules like DNA. *Transformation* pokes "holes" on the cell membrane to allow DNA entry. Two methods are available for transformation: heat shock and electroporation. Heat shock involves quick heating and cooling of bacterial cells that were previously treated with $CaCl_2$. As cell membranes are permeable to Cl⁻ ions, once Cl⁻ goes through as a hydrated anion, it leads to "swelling of the cell". The 42 °C heat shock leads to formation of holes (pores) on the membrane that allows DNA entry. During electroporation, these holes are generated via a brief electroshock.

Transformation is a low-yield procedure: relatively few cells take up the plasmid, despite the heat shock or electroporation treatment.

5.5 Cell Growth

As a result of the low yield, a selective marker has to be used to distinguish cells that harbor the plasmid from cells that do not. The most commonly used selective marker with bacterial systems is the usage of an antibiotic. Plasmids harbor a resistance gene for a given antibiotic. This means cells that took up the plasmid will grow on media containing the antibiotic, while cells that did not take up the plasmid will not. Consequently, plating the transformation mixture on solid medium that contains the selective antibiotic, ampicillin in our case, will only yield colonies from bacteria that harbor the plasmid containing the ykkCD sensor RNA.

Once *E coli* cells that harbor the plasmid DNA containing the mutant ykkCD sensor are identified, these cells are inoculated into liquid medium containing the selective antibiotic (ampicillin) and grown for 16 hrs. at 37 °C with vigorous shaking.

5.6 Purification of Plasmid DNA from Bacterial Cells

When plasmid DNA is synthesized via PCR amplification, the resulting DNA is very clean. However, when DNA is synthesized using bacterial cells the plasmid DNA is contaminated with bacterial cell components: their cell membrane, cellular proteins, DNA, and RNA. During the plasmid prep, all of these contaminants have to be removed. Plasmid preps utilize the fact that DNA is highly negatively charged (compared to proteins), very soluble in water (compared to lipids and membrane proteins) and fairly small (compared to genomic DNA).

Make sure you follow the plasmid prep protocol exactly as written. The first few steps of plasmid purification (from cell breakage to loading the DNA onto the purification column) significantly impact the purity and quantity of the plasmid DNA.

The exact protocol of plasmid purification is dependent of the kit used and will be provided in lab. A general outline of procedure is presented below.

1. Plasmid preps began with breaking the bacterial cell membrane (lysis) to free the plasmid DNA. This step is usually achieved by an alkaline lysis. Cell breakage can also be achieved using the enzyme lysozyme. *Be careful to NOT allow cell lysis to proceed longer than instructed. Handle cell lysate gently to prevent shearing of chromosomal DNA. This could cause contamination of the plasmid DNA with genomic DNA.*

2. After cell lysis, the plasmid DNA will be in solution with soluble proteins, but anything that is not water-soluble, the genomic DNA, lipids and membrane proteins, will form sediment. This sediment can either be removed by centrifugation or filtration. Either way, this step purifies the plasmid DNA from proteins, lipids, and genomic DNA.

3. The plasmid DNA is further purified via column purification. Often a silica-based anion-exchange chromatography column is used. This step takes advantage of the negative charge of DNA to separate the plasmid DNA from the host's genomic DNA. Since genomic DNA is much larger it has a higher negative charge then plasmid DNA, it interacts more strongly with the anion exchange column. As a result genomic DNA remains bound to the anion exchange column while the plasmid DNA is eluted.

5.7 Agarose Gel Electrophoresis

To evaluate the quality of the plasmid DNA agarose gel electrophoresis will be used. Electrophoresis is a technique that is used to separate molecules based on charge and shape. Separation is achieved by moving the negatively charged molecules like DNA through an electric field from the negative to the positive electrode. Shorter molecules migrate faster than longer ones. To prevent spreading of the sample during electrophoresis a solid medium is used like agarose or polyacrylamide. *Agarose* is a polysaccharide that is generally extracted from seaweed and is used to separate large nucleic acids (> 500 base pair). The pore size of the polymer is dependent on the percentage of the agarose used (0.6 – 2%). Higher percentage leads to smaller pore size and is designed to separate smaller molecules. Agarose gels generally do not denature (unfold) DNA. Therefore, DNA conformation also affects the rate of migration.

The rate of migration in an agarose gel is described by the equation below where *v* is velocity (rate), *q* is the charge of the molecule, *E* is the strength of the electric field and *f* is friction.

$$v = q \, E \, / \, f \tag{1}$$

DNA migrating through a gel with smaller pore size experiences more friction and travels less distance than the same molecule migrating through a gel with larger pore size. Likewise, DNA with more elongated shape travels smaller distance in the gel than compact DNA, because compact DNA experiences less friction as it migrates through the gel. Therefore, supercoiled DNA (more compact) migrates faster in an agarose gel than nicked or circular DNA (more elongated shape). DNA with higher molecular weight travels less far than DNA with lower molecular weight, as long as their shape is similar. The relationship between shape, molecular weight and distance traveled in an agarose gel is seen in *Fig. 5.2.*

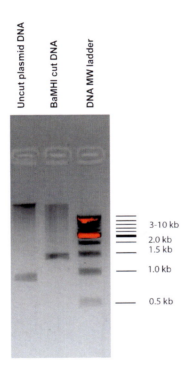

Figure 5.2: Migration of DNA on an agarose gel depends on shape and size. Shorter DNAs migrate faster than longer DNAs. Note the separation of DNAs in the MW marker.

Nucleic acids are visualized on an agarose gel using ethidium bromide (EtBr) staining. EtBr is fluorescent and it interacts with the nucleobases in DNA (intercalates with DNA bases). Excess of the EtBr migrates out of the gel quickly due to its small size. When exposed to UV light nucleic acids "light up" in an agarose gel due to EtBr fluorescence. SYBR green nucleic acid stains work in a similar fashion (*Fig 5.3*).

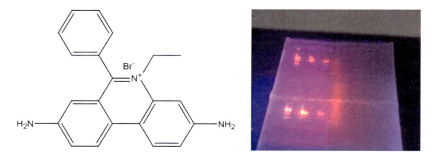

Figure 5.3: Nucleic acid visualization using EtBr: EtBr is aromatic. As a result, it chelates with nucleobases (left). EtBr treated nucleic acids light up when shined with UV light, because EtBr is fluorescent in the ultraviolet range (right).

5.8 Application of Agarose Gel Electrophoresis

Agarose gel electrophoresis is a powerful technique that is used to separate nucleic acids by size and/or shape. It has both preparative and analytical applications. As an analytical tool, agarose gel electrophoresis can be used to evaluate the success of any enzymatic reaction that results in significant change of DNA shape, such as cleavage by restriction endonucleases (see more detail in Chapter 6). Restriction endonucleases are enzymes that cleave DNA, and thus turn supercoiled DNA (fast migrating) into linear DNA (slow migrating). As a result, the success of a cleavage reaction by restriction endonuclease can be judged by how far the cleaved DNA sample migrates on an agarose gel as compared to the original (uncut) DNA. *Fig 5.2* compare lanes with cut and uncut DNAs. A preparative application of agarose gel electrophoresis is to remove any uncut (supercoiled) DNA from a sample treated with a restriction endonuclease. In this case, the slow migrating band (cut DNA) is excised from the agarose gel and the DNA is eluted from the gel piece using commercially available kits.

5.9 DNA Quality Control Using UV Spectroscopy

UV spectroscopy is commonly used to determine the concentration of plasmid DNA. Nucleic acids strongly absorb UV light at 260 nm, due to the nucleobases (heterocyclic aromatic compounds). The concentration of DNA produced is estimated using the formula below:

$$\text{Beer-Lambert law} \quad \text{Absorbance (at 260nm)} = e^\star c^\star l \quad\quad\quad (2)$$

Where e is the absorption coefficient (20 ml/mg*cm for plasmid DNA), l is the path length (1 cm) and c is the concentration of the plasmid DNA in mg/ml. By measuring the absorbance of the plasmid DNA produced, the concentration can be calculated.

PROTOCOLS

Reagents and equipment needs are calculated per six student teams. There is ~20% excess included.

Equipment/glassware needed
  1. Three 100 ml corked Erlenmeyer flask
  2. Three 100 ml graduated cylinder
  3. Three sets of micropipettes 20-100 µl and 2-20 µl.
  4. UV spectrometer
  5. Gel documentation system
  6. Water bath (for transformation)
  7. Orbital shaker (for cell growth)
  8. Six 10-15 ml tubes (sterile)
  9. Six 1.5 ml centrifuge tubes (Eppendorf)

Solutions needed
  1. 3 l 1 x TAE (Tris-acetate-EDTA) pH=8.0
  2. 30 ml 1% EtBr (or equivalent nucleic acid stain)
  3. 20 ml Agarose gel loading dye (6x, Biorad)
  4. 30 ml 1 kb molecular weight ladder (Biorad)
  5. 30 ml LB (Luria-Bertani) broth and six LB/Ampicillin (100 mg/ml) plates
  6. 30 ml 100 mg/ml (1000 x) Ampicillin (for cell growth)

Transformation (may be done by teaching assistants ahead of time)
  1. Set water bath to 42 °C.
  2. Label 1 sterile glass or plastic tube and 1 microcentrifuge tube for each mutant.
  3. Label 1 LB-agar plate containing 100 mg/ml ampicillin for each mutant.
  4. Thaw *E. coli* Dh5α competent cells on ice (100 ml competent cell is used for each mutant).
  5. Add 10 µl PCR amplified mutant DNA to each tube containing competent cells. Keep cells on ice for 20 min.
  6. Heat shock DNA as follows:
      a. Place cell into 42 °C for 2 min.
      b. Move cells to ice for 3 min.

7. While performing the heat shock treatment pipette 900 µl LB broth into previously labeled sterile tubes and place them into 37 °C orbital shaker for pre-warming.
8. Add heat shocked cells to tubes containing 900 µl pre-warmed LB broth.
9. Grow cells while shaking gently (120 rpm) for 1 hr.
10. Transfer cells into previously labeled centrifuge tubes. Pellet cells for 2 min in a microcentrifuge using 5000 rpm.
11. Remove supernatant. Resuspend cells in 100 µl LB broth and plate then on the previously labeled LB/ampicillin plates.
12. Grow cells overnight in 37 °C incubator.

Cell growth (may be done by teaching assistants ahead of time)
1. Label 1 sterile glass tube for each mutant.
2. Inoculate a single colony into 1 ml LB/ broth supplemented with 100 µg/ml ampicillin. Grow cells for 5-6 hrs. in an orbital shaker at 37 °C with vigorous shaking (280 rpm).
3. Label 1 sterile 120 ml Erlenmeyer flask for each mutant.
4. Use 100 µl cells from "step 2" to inoculate 40 ml LB broth supplemented with 100 µg/ml ampicillin for each mutant.
5. Grow these cells overnight with vigorous shaking (280 rpm).
6. Cells may be harvested (10 min centrifugation with 5000 x g; remove supernatant) and stored at -20 °C.

Plasmid purification from E. coli cells
Protocol depends on kit used and will be provided in lab. See pg 47 for a basic overview.

Agarose gel electrophoresis
1. Weight 1 g agarose and place it to a corked Erlenmeyer flask provided.
2. Add 100 ml 1 x TAE buffer using graduated cylinder.
3. Heat for 1 min in microwave, swirl flask and briefly cool under water.
4. Add 10 µl EtBr (Wear gloves. EtBr is mutagenic); swirl flask for mixing.
5. Pour liquid into gel cassette and position combs. It takes about 20 min for the agarose gel to set.
6. Once gel is set load your sample plus 10 µl molecular weight ladder (with 2 µl loading dye). Run gel for 20 min with 100 V.
7. Take picture of gel using a gel documentation system.

UV spectroscopy (this protocol is for Nanodrop spectrometer; for conventional UV spectrometer large volumes of DNA need to be used)
1. Turn on spectrometer and choose nucleic acid assay setup.
2. Blank spectrometer with 2 µl Millipore water.

3. Measure the absorbance of a water sample to ensure that the spectrometer is clean.
4. Use 2 µl plasmid DNA sample to measure the concentration of your plasmid DNA.
5. Record DNA concentration in mg/ml.

Notes to the instructor

The experiment in Chapter 5 is designed to isolate plasmid DNA containing mutant ykkCD tetracycline sensor RNA using *E. coli* cells. The same protocol with minimal modification could be used to perform plasmid purification of any DNA from *E. coli* cells. At the authors' institution, students were provided with liquid cell cultures that contained the mutant toxin sensor DNA. Transformation and cell growth was performed by teaching assistants in the absence of students. This setup saves lab time, but somewhat limits students' understanding of the steps required to make plasmid DNA. If desired, a separate lab session may be allocated where students perform transformation and cell growth. Any commercially available *E. coli* Dh5α competent cell is suitable for transformation. The *E. coli* strain and the protocol to make them competent for transformation used by the authors is available upon request. Plasmid Midi Kit from Qiagen works well to purify plasmid DNA, but similar kits from other vendors may work well. A mid-size plasmid purification kit is recommended to ensure that sufficient quantity of the plasmid DNA is produced.

Prelab Questions for Plasmid Prep

Define the following terms.

1. Transformation

/ 1 pts

2. Agarose gel

/ 2 pts

3. Supercoiled DNA vs Linear DNA

/ 2 pts

4. Beer-Lamber law. Define each term in the equation.

/2 pts

5. What type of precautions do you have to observe while handling EtBr and why?

/ 3 pts

DNA Purification

Lab Report Outline and Point Distribution

1. Several sentences defining the goal/purpose of this experiment. Please indicate the role this step plays in the mini project (5 pts.).

2. Briefly describe the strategy used to purify plasmid DNA from cells (10 pts.). (What are the major contaminants? What properties of your plasmid DNA are exploited during purification and how? What are the differences between your plasmid DNA and other cellular macromolecules?)

3. Draw a detailed flowchart of the plasmid prep. Make sure you indicate the volume of each buffer that needs to be added and which fraction (soluble/insoluble or column/filtrate) contains the plasmid DNA and which fraction is being discarded (20 pts.).

4. Define/describe transformation (5 pts.).

5. Define/describe the role the DpnI digestion plays in the Quickchange process (5 pts.).

Troubleshooting a plasmid prep
6. You colleague had significant genomic DNA contamination in their plasmid prep. In your opinion at which step(s) of the preparation a mistake was made? Briefly explain (5 pts.).

Electrophoresis Problem Set

1. (9 pts.) Describe the effects the following would have on a DNA agarose gel electrophoresis. Use the equation, $v = qE/f$, in a qualitative way as part of your explanation.

   a. The agarose concentration is increased from 1% to 1.5%.

   b. The voltage for the electrophoresis is decreased from 100 volts to 60 volts.

   c. The pH was changed from about neutral to acidic conditions (pH=3.0).

2. (5 pts.) You are asked to separate 1 μg of DNA on an agarose gel. The sample is to contain 1 x TAE buffer and the sample size is 10 μl. What volumes of the following would you use to prepare this sample: 0.5 μg/μl DNA stock solution, 10 x TAE buffer and water.

3. (9 pts.) An agarose gel was run to determine the size of several pieces of DNA. Lane "M" has a DNA size ladder of 11 steps (top 10 kb, 9 kb, 8 kb, 7 kb, 6 kb, 5 kb, 4 kb, 3 kb, 2 kb, 1 kb, 0.5 kb bottom). Measure the distances traveled to the nearest millimeter. Construct a plot of the log(kb) versus the distance traveled for the DNA ladder.

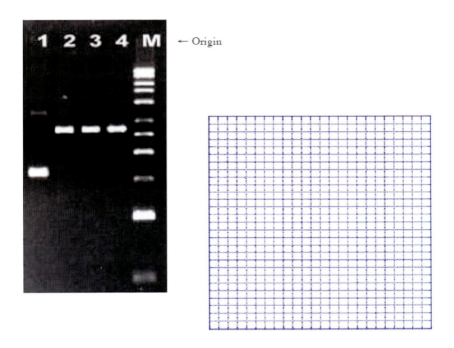

(a) Using the plot, determine the size of the DNA (in kb) in lane 4 and lane 1.

(b) The large band in Lane 1 is actually a circular plasmid DNA of the same size as the linear DNA in Lane 4. Briefly explain why the circular plasmid travels further in the gel than the linear DNA.

6 Preparing DNA Template for Mutant RNA Sensor Synthesis Using a Restriction Endonuclease

6.1 Learning Objective

In this lab, you get our DNA template containing the mutant sensor RNA ready for RNA synthesis. With this process, you will learn how restriction endonucleases – enzymes that cleave DNA at a given sequence – achieve their extraordinary specificity.

6.2 Mini Project Flowchart

The bolded block in the flowchart below highlights the role of the current experiment in the mini project.

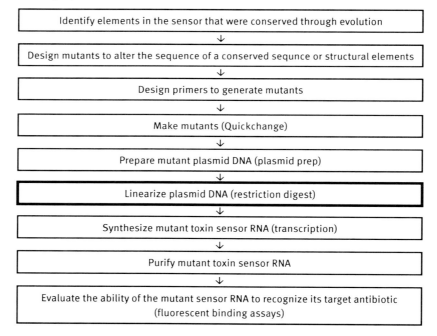

6.3 Synopsis

The goal of the mini project was to identify nucleotides in the toxin sensor that are essential for recognizing the antibiotic tetracycline. In the quest toward this goal, you identified evolutionarily conserved elements in the sensor and modified them. You designed primers to change the sequence of the sensor and made mutant sensor DNAs using PCR amplification and plasmid preparation. To test whether these mutant sensors are capable of recognizing tetracycline you have to synthesize the mutant sensor RNAs. Recall, the sensor is an RNA molecule, but its sequence is coded in the genome at the DNA level. RNA is synthesized using a linear DNA template and an RNA polymerase. Therefore, the plasmid DNA that codes for the mutant sensor has to be cut (linearized) using a restriction endonuclease. The plasmid vector is cleaved at the end of the toxin sensor sequence to prevent the RNA polymerase from transcribing the entire plasmid vector into RNA (you only need the sensor). RNA polymerases work similarly to DNA polymerases and you will learn more about them in detail during the next lab.

6.4 How do Restriction Endonucleases Work?

Restriction endonuclease evolved to cleave intruder DNA, but leave the DNA of the host organism intact. They are used extensively by molecular biologists as a tool to specifically cleave plasmid or *in vitro* synthesized DNA during cloning. Restriction endonucleases are enzymes that cut dsDNA at a specific sequence with million-fold specificity. This means that restriction endonucleases favor their target sequence million-fold over other sequences. *Cognate DNAs* have the specific sequence and are cleaved efficiently whereas *non-cognate DNAs* do not have the specific sequence, and therefore are not cleaved efficiently by the restriction enzyme. Restriction enzymes use a metal-activated water molecule as nucleophile. The metal-bound water molecule is more acidic than a regular water molecule. Therefore, the metal-bound water loses its proton more easily and becomes a hydroxyl group. A hydroxyl group is a strong nucleophile and able to catalyze cleavage of the phosphodiester bond very efficiently (*Fig. 6.1*).

Figure 6.1: Mechanism of action of restriction endonucleases: Restriction endonucleases cleave dsDNA with high specificity. They use a metal-ion bound water molecule as nucleophile to catalyze cleavage of the phosphodiester bond.

6.5 How do Restriction Enzymes Achieve Million-Fold Specificity?

Restriction endonucleases favor their cognate sequence million fold over other sequences. How is this extraordinary specificity achieved? Restriction endonuclease cleavage sites have twofold symmetry. Likewise, restriction enzymes are dimers and have complementary twofold symmetry (*Fig. 6.2*).

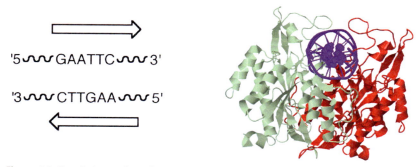

Figure 6.2: Restriction endonucleases and their target sites have complementary twofold symmetry. The monomers of the restriction endonuclease homodimer are shown in green and red; the dsDNA template is purple (PDB ID 1ERI).

The restriction endonuclease forms a more extensive H-bonding network with its cognate than with its non-cognate DNA. As a result, the cognate DNA is bent (it has a kink in the dsDNA structure) and the *scissile bond* (the bond that is cleaved) is placed right in the middle of the enzyme's active site for cleavage. The energy required to bend the cognate DNA comes from the extensive H-bonding between restriction enzyme and the cognate DNA (*Fig. 6.3; Fig 6.4*).

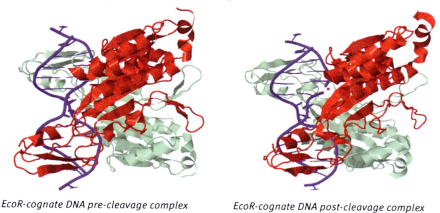

EcoR-cognate DNA pre-cleavage complex
(PDB ID 1ERI)

EcoR-cognate DNA post-cleavage complex
(PDB ID 1QPS)

Figure 6.3: The cognate DNA is bent by the restriction endonuclease to position the scissile bond into the active site for cleavage.

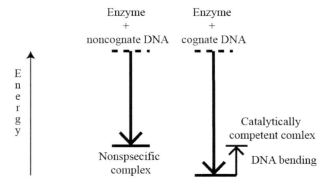

Figure 6.4: The extensive H-bonding between the restriction endonuclease and its cognate DNA substrate provides the energy for DNA bending. DNA bending is essential for catalysis (see text).

6.6 How Do We Judge Whether The Plasmid DNA is Successfully Linearized?

Agarose gel electrophoresis is a powerful technique to separate DNA based on size and/or shape (see Chapter 5.). Since plasmid DNA coding for the mutant sensor RNA was synthesized by bacteria, it is supercoiled and compact. Once this DNA is cut with a restriction endonuclease, it is nicked and hence has a more elongated shape. As a result supercoiled (uncut) plasmid DNA migrates farther in the agarose gel than linearized (cut) DNA. If the DNA linearization is successful, the sample containing the cut DNA should have a single band that migrates less far in the gel than the uncut DNA sample (*Fig. 6.5*).

6.7 What are We Going to do in the Lab?

First, you will digest the plasmid DNA coding for the mutant toxin sensor with the restriction enzyme BamHI and evaluate the success of the reaction using agarose gel electrophoresis. While the gel is running, you will remove the restriction enzyme by phenol-chloroform extraction and increase DNA concentration by ethanol precipitation.

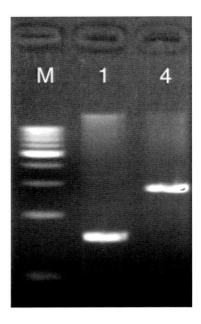

Figure 6.5: The success of restriction endonuclease cleavage is judged by resolving the reaction mix on an agarose gel.

PROTOCOLS

Reagents and equipment needs are calculated per six student teams. There is ~20% excess included.

Equipment/glassware needed

1. Three 100 ml corked Erlenmeyer flask
2. Three 100 ml graduated cylinder
3. Three sets of micropipettes (20-100 μl and 2-20 μl)
4. Gel documentation system
5. Water bath (for restriction digest) set to 37 °C
6. Thirty 1.5 ml centrifuge tubes (Eppendorf)
7. 1 microcentrifuge capable of 10,000g

Solutions needed

1. 3 l 1 x TAE (Tris-acetate-EDTA) pH=8.0
2. 30 ml 1% EtBr or equivalent nucleic acid stain
3. 20 ml agarose gel loading dye (6x, Biorad)
4. 30 ml 1 kb molecular weight ladder (Biorad)
5. 2 ml Phenol:Chlorophorm:isoamyl alcohol mix

6. 150 ml 3M Na-acetate, pH=5.2
7. 1.5 ml isopropanol

Restriction digestion of pUC19-ykkCD plasmid vector
Each member of the team performs a digestion

Mix the following solutions in a centrifuge tube
   1. 5 µl buffer 3 (New England Biolabs)
   2. 5 µl 1 mg/ml bovine serum albumin solution
   3. 5 µl BamHI restriction enzyme (New England Biolabs)
   4. 75 µl plasmid DNA

Incubate in 37 °C water bath for 1 hr.

Agarose gel electrophoresis
   1. Weight 1 g agarose and place it to the corked Erlenmeyer flask provided.
   2. Add 100 ml 1xTAE buffer using a graduated cylinder.
   3. Heat for 1 min in microwave, swirl flask and briefly cool under water.
   4. Add 10 µl EtBr. (Wear gloves. EtBr is mutagenic.) Swirl flask for mixing.
   5. Pour liquid into gel cassette and position combs. It takes about 20 min for an agarose gel to set.
   6. Once the gel is set load your samples with a molecular weight ladder. Gel should be run for 20 min with 100 V.
   7. Take picture of the gel using a gel documentation system.

Load the 5 ml of the following samples mixed with 1ml loading dye onto the agarose gel
   1. Molecular weight marker
   2. Plasmid DNA from the plasmid prep (uncut DNA sample)
   3. Plasmid DNA digested with the restriction endonucleases (cut DNA)

Phenol-chloroform extraction and ethanol precipitation
   1. Add 100 µl phenol/chloroform mixture to your sample.
   2. Vortex for 20 sec.
   3. Spin for 2 min in a microcentrifuge at maximum speed.
   4. Discard bottom layer. (This step removes the organic layer containing the restriction endonuclease.)
   5. Add 100 µl chloroform, vortex and spin as before. Remove bottom layer. This step removes residual phenol.
   6. Add 10 µl 3 M NaAc and 100 µl isopropanol to your mixture.
   7. Vortex sample and place into -20 ºC freezer for precipitation.

Notes to the instructor

The experiment in Chapter 6 is designed to teach students how to perform enzymatic cleavage of the pUC19-ykkCD plasmid vector using BamHI restriction enzyme. The same protocol with minimal modifications may be used with a different DNA and/or restriction endonuclease. BamHI restriction enzyme was purchased from New England Biolabs, but other vendors may be used. Two student teams shared one agarose gel to limit equipment need while maximizing exposure to this important technique. If the number of equipment is limited six student teams can easily share one agarose gel apparatus. Since restriction endonuclease digest takes an hour it is convenient to start the laboratory session by setting up the reactions and giving pre laboratory lecture while the reactions are running. This laboratory has two breaks while the agarose gel sets and runs. These times can be utilized to complete either the restriction endonuclease or electrophoresis worksheet.

Prelab Questions for DNA Linearization

Define the following terms.

1. Restriction endonuclease

/ 1 pts

2. Scissile bond

/ 1 pts

3. Cognate DNA and noncognate DNA

/ 2 pts

4. To test DNA quality you need to resolve the DNA sample on an agarose gel. You need 1 l 1 x TAE buffer (Tris-acetate EDTA) to make and run an agarose gel. You purchased 40 x concentrated TAE buffer stock. Describe how you make the 1 l 1 x TAE solution needed for the gel (show your calculation).

/ 4 pts

DNA Linearization
Lab Report Outline and Point Distribution

1. Several sentences defining the goal/purpose of this experiment. Describe the significance of this experiment within the mini project. (3 pts.)

2. Briefly define/describe a restriction endonuclease. Explain how specificity is achieved. (4 pts.)

3. Include a copy of your agarose gel scan with your samples marked. (4 pts.)

4. Report the concentration of your plasmid DNA. (2 pts.)

5. Evaluate plasmid DNA yield based on the DNA concentration (how many mg of DNA did you get). Show your yield calculation and comment on how good was your DNA yield. (6 pts.)

6. Evaluate plasmid DNA purity based on the agarose gel scan and the 260/280 reading determined by UV absorbance. Explain. (6 pts.)

7. Judge the success of your DNA linearization based on the agarose gel scan. Explain. (4 pts.)

8. Judge the size of your plasmid DNA provided that the DNA step ladder has sizes: 1000 bp, 2000 bp, 3000 bp, 4000 bp, 5000 bp, 6000 bp, 7000 bp, 8000 bp, 9000 bp, 10,000 bp. Explain. (4 pts.)

9. Electrophoresis Problem Set (17 pts.)

Worksheet - Restriction Endonucleases

Restriction endonucleases are very specific and cleave DNA at specific nucleotide sequences. When these enzymes act on large DNA, it is broken down into a set of many, many fragments of different sizes. The size distribution of such fragments is essentially unique for each different DNA. This distribution provides a DNA "fingerprint" that is unique for each individual. DNA fingerprints have been used to identify genetic mutations, to trace hereditable traits, to establish parentage, and to place suspects at the scene of a crime.

There are several different techniques that give rise to a DNA fingerprint. Some of the problems below will describe techniques and ask you to interpret the results.

1. (3 pts.) There are thousands of different restriction endonucleases, each with its own specificity. For example, BamHI, EcoRI, and XhoI have the following specificities (the cleavage sites are marked with arrows):

$$
\text{BamHI} \quad \begin{array}{l} 5'\ \text{G} \downarrow \text{GATCC}\ 3' \\ 3'\ \text{CCTAG} \uparrow \text{G}\ 5' \end{array}
\qquad
\text{EcoRI} \quad \begin{array}{l} 5'\ \text{G} \downarrow \text{AATTC}\ 3' \\ 3'\ \text{CTTAA} \uparrow \text{G}\ 5' \end{array}
\qquad
\text{XhoI} \quad \begin{array}{l} 5'\ \text{C} \downarrow \text{TCGAG}\ 3' \\ 3'\ \text{GAGCT} \uparrow \text{C}\ 5' \end{array}
$$

Identify the cleavage sites on the following DNA (mark each site with a line and the name of the specific endonuclease):

5′ CCCGAGGATCCTTAGGAATTCATCTA 3′
3′ GGGCTCCTAGGAATCCTTAAGTAGAT 5′

2. (4 pts.) Show the sequence of the product fragments from the BamHI cleavage. Why are these products said to have "sticky ends?"

3. (6 pts.) Restriction fragment length polymorphism (RFLP, often pronounced "riflip") is the oldest form of DNA fingerprinting. It depends on the fact that the DNA from each individual has many segments with unique sequences. Each segment can have a variety of sequences that are inheritable. This characteristic is termed "polymorphism." Polymorphism means that each individual have DNA with different restriction enzyme cleavage sites. Briefly, RFLP involves
   a. Treating the DNA with a specific restriction endonuclease;
   b. Separating the resulting fragments by size on an agarose gel;
   c. Visualizing specific fragments using complementary DNA probes.

The following agarose gel was used to establish parentage. It was generated using single locus RFLP. That is the probe used was sensitive to only one polymorphic segment in the DNA.

To interpret these results you need to know
   – Each individual carries two copies of DNA and, meaning, there are two different polymorphisms for each DNA segment;
   – Each individual inherits one DNA copy from each parent.

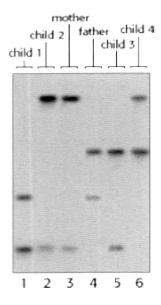

(Courtesy of The Biology Project - University of Arizona: http://www.biology.arizona.edu/)

a. What must be true of the child's DNA fingerprint in order to establish that he/she is related to the mother and father?

b. Are there any children that don't appear to be biological children of these parents? Explain.

4. (4 pts.) In the following DNA fingerprints, C/AF Mix is a lane that contains both the alleged father's DNA as well as the child's DNA.

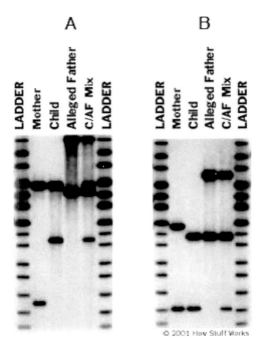

(Courtesy of Genelex Corporation)

Is paternity proven in either case? Briefly explain.

5. (6 pts.) PCR has been combined with RFLP to increase the sensitivity of the DNA fingerprinting. In this technique PCR amplifies a specific DNA segment of interest. Then, changes in the sequence are identified by RFLP.

For example, sickle cell anemia can be diagnosed using this technique. Sickle cell anemia is caused by a single mutation on the DNA that code for one of hemoglobin protein chains. This mutation also eliminates a restriction endonuclease cleavage site for the enzyme; Cvn. Loss of the cleavage site is then diagnostic for the sickle cell gene.

In the following, a mother, father and children have been tested for the sickle cell gene via DNA fingerprinting. The diagram above the electrophoresis pattern shows parents (row I) and children (row II). Females are symbolized by a circle and males by a square. The normal gene is indicated with an open symbol and the sickle cell gene by the filled symbol. (Remember each individual carries two copies of the gene.)

Each individual's symbol corresponds to a lane in the electrophoresis gel. The lanes are also marked with "S" for presence of the sickle cell mutation and "A" for presence of the normal gene. (The DNA ladder is on the right and is given in number of bases. Note that the gel is "upside down" - the smaller fragments are closer to the top.)

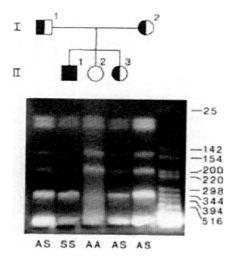

(Courtesy of Dr. Carole Ober)

a. Estimate the size of the cleavage fragments that are only present in the individual without the sickle cell mutation. Briefly explain your choice.

b. From what band do these cleavage fragments come? Can you identify the band and estimate its size? Explain your reasoning.

c. The researcher diagnosed these subjects (SS, AS, or AA) based on this electrophoresis. Do you see any inconsistencies in the diagnosis? Briefly explain.

Cloning Experiment Design - Worksheet

Design primers for conventional (cutting and pasting) mutagenesis to place the ykkCD toxin sensor into the cloning vector pUC19. Experiments like this are usually the first step in studying a protein or RNA in the lab. The sequence of the toxin sensor and the map of the cloning vector are given below.

DNA sequence of ykkCD sensor:

    Tgtaaagttttctagggttccgcatgtcaattgacatggactggtccgagagaaaacacatacgcgtaaataga agcgcgtatgcacacggagggaaaaaagcccgggagag

Map of the pUC19 cloning vector

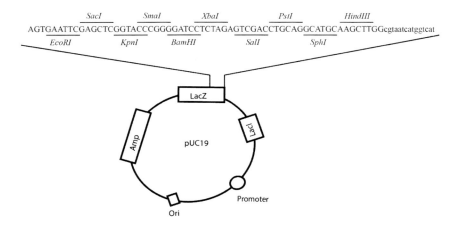

The ykkCD toxin sensor has to be inserted into pUC19 using a part of the cloning vector called multiple cloning site or polylinker MCS. This part of the cloning vector contains cut sites for several restriction enzymes and designed to allow insertion of DNA sequence. In case of pUC19 the available restriction enzymes are: EcoRI, SacI, KpnI, SmaI, BamHI etc.

To complete the task of pasting the ykkCD sequence into pUC19 answer the questions below:

1. Step one in a conventional cloning project is to choose a pair of restriction enzymes from the ones in the polylinker MCS that do not cleave the sequence of interest (the ykkCD sensor). Restriction enzymes that cut the ykkCD sequence cannot be used for cloning, because they would cleave the sequence of interest and would not allow pasting the entire sequence into the cloning vector. To select appropriate restriction endonucleases follow the two steps below.

a. Paste the sequence of the ykkCD sensor into a program called Webcutter (http://www.firstmarket.com/cutter/cut2.html). List the names of restriction enzymes that do not cut the ykkCD sequence.

b. Which enzymes in the MCS polylinker may be used to perform your cloning project? Provide a brief explanation.

2. Design primers to perform the cloning experiment.

a. Provide the top and bottom primer sequences. Briefly explain how you designed these primers. Which part of the DNA these primers anneal to? Remember, there is a difference between Quickchange and conventional PCR.

b. Calculate the T_M (melting temperature) for each primer. You may use Oligo calculator (http://www.basic.northwestern.edu/biotools/oligocalc.html) to predict T_M.

c. Conventional PCR requires addition of the special sequence that is recognized by each restriction enzyme called restriction site. You may find the recognition sequence for each restriction enzyme at http://www.neb.com.

Your revised primer sequences should follow this pattern

Restriction site sequence + primer sequence from "b".

d. Write the sequence of both primers in the 5' to 3' direction.

7 Synthesizing the ykkCD Mutant Toxin Sensor RNA *in vitro*

7.1 Learning Objective

In the quest toward understanding how the ykkCD toxin sensor recognizes the antibiotic tetracycline you thus far designed mutants to alter the sequence of the sensor, and made the plasmid vectors containing the mutant sensor using PCR amplification. You purified these plasmids from bacterial cells and prepared them to be templates for RNA synthesis. In this lab you will learn how RNA polymerases work. You will synthesize the mutant sensor *in vitro* using the plasmid DNA template and T7 RNA polymerase.

7.2 Mini Project Flowchart

The bolded block in the flowchart below highlights the role of the current experiment in the mini project.

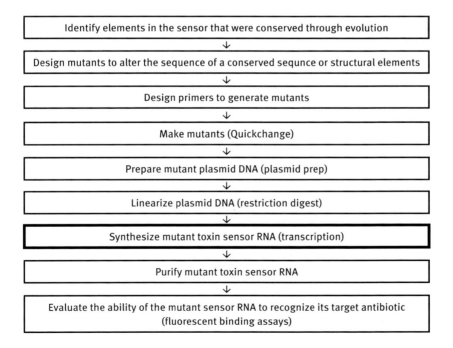

7.3 How do RNA Polymerases Work?

RNA polymerases are enzymes that synthesize RNA using a DNA template. RNA synthesis using RNA polymerases is also called transcription. During transcription we copy sequence information coded in the genome (DNA) to RNA. When the sequence codes for protein the resulting RNA sequence is used as template for protein synthesis. The chemistry of RNA polymerization is identical to that of DNA synthesis (Chapter 4). The 3'-OH group of the growing polynucleotide chain acts as a nucleophile to attack the phosphorous of the incoming nucleotide triphosphate (NTP) thereby forming the phosphodiester bond. Divalent metal ions (Mg^{2+}) are required for transcription to position the 3'-OH group for nucleophilic attack (*Fig. 7.1*).

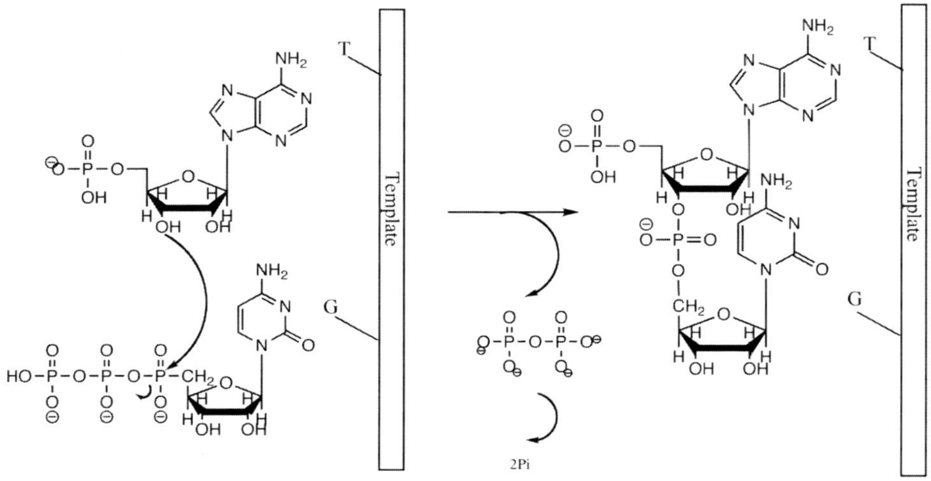

Figure 7.1: Chemistry of polymerization. The 3'-OH group of the growing nucleotide chain acts as nucleophile to attack the α-phosphorous of the incoming NTP to catalyze formation of the phosphodiester bond.

During polymerization, a high-energy bond (the anhydride bond in the incoming NTP) is broken and a high-energy bond (the phosphodiester bond) is formed. What makes polymerization energetically favorable? The breaking of the pyrophosphate (PP_i) into two inorganic phosphates is what drives polymerization. Without this step, polymerization would not be energetically favorable. During transcription the polymerase first unwinds the DNA template, then recruits the nucleotide triphosphate (NTP) that is complementary to the template and finally it catalyzes formation of the phosphodiester bond. The polymerase moves from 3' to 5' direction on the template DNA while synthesizing RNA. The schematic depiction of this process is also called the transcription bubble (*Fig. 7.2*).

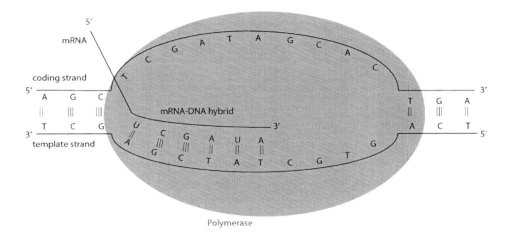

Figure 7.2: The transcription bubble: After separating the dsDNA template the RNA polymerase moves from 3' to 5' direction on the DNA template to synthesize RNA. The RNA sequence is the same as the top DNA strand, but each T is replaced with U.

7.4 How Does Transcription Start?

Polymerases are recruited to the DNA template by specific sequences called promoters (*Fig 7.3*). Each polymerase has its own promoter sequence. Once the polymerase encounters this sequence it binds to the DNA template and initiates RNA synthesis.

Figure 7.3: Schematics of a bacterial promoter. Note that the two specific sequences are required to be positioned at a given distance from each other for optimal polymerase binding.

Polymerases are oligomer enzymes, meaning each polymerase has a catalytic domain that facilitates the chemistry of polymerization. They also have a domain that recognizes the promoter and helps recruit the polymerase to the DNA template. In *E. coli* this domain is called the sigma factor (*Fig. 7.4*).

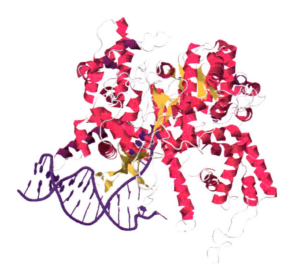

Figure 7.4: T7 RNA polymerase transcription initiation complex (PDB ID 1QLN): The T7 RNA polymerase recognizes a hairpin-shaped promoter. The RNA chain grows in the 5' to 3' direction. The RNA sequence is the same as the top DNA strand except each T is replaced with U.

7.5 How Does Transcription End?

In bacteria transcription ends when the polymerase encounters a stable structure that it cannot unwind. This stable structure is also called the *terminator* or *terminator stem*. The polymerase stalls when encountering a stable structure, and together with the nascent RNA, it is released from the DNA template thereby ending transcription (*Fig. 7.5*).

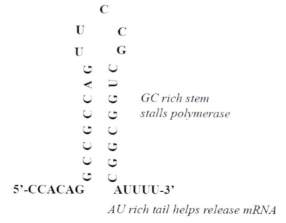

Figure 7.5: Schematics of a terminator stem: The G/C rich stem is stable and stalls the polymerase; the A/U rich tail releases the nascent RNA.

In research laboratories, scientists most commonly use T7 RNA polymerase for *in vitro* transcription, because it can synthesize milligram amounts of RNA within two hours. The T7 RNA polymerase is from the T7 bacteriophage. The promoter of T7 RNA polymerase is 19-nucleotide long and folds into a hairpin. This sequence has to be present in the cloning vector upstream of the DNA template sequence to enable successful transcription. Since the T7 RNA polymerase is very robust, most terminator stems are not stable enough to stall it. Therefore, to terminate transcription at the end of the desired sequence, the template DNA must be cut (linearized) prior to RNA synthesis to stop the T7 RNA polymerase. Meaning, the polymerase falls off the DNA template at the cut site thereby ending transcription. In case of the ykkCD sensor RNA, the cleavage site for the BamHI restriction endonuclease directly follows the ykkCD sequence in the cloning vector. This site recruits the BamHI restriction enzyme that cleaves (linearizes) the template DNA prior to transcription (Chapter 6.)

7.6 Transcription in Practice

Several kits are commercially available to perform *in vitro* transcription with high yield. These kits contain T7 RNA polymerase (enzyme), NTPs and a reaction buffer. The reaction buffer has pH~8.0 and contains magnesium ions (usually in the form of $MgCl_2$). Transcription is performed in a 37 °C water bath for 1 or 2 hrs. Inclusion of a ribonuclease (RNase) inhibitor often increases the yield and quality of *in vitro* synthesized RNA, because it inhibits ribonucleases that might degrade the synthesized RNA.

7.7 What Are We Going To Do Today?

Today we will precipitate the DNA template, check DNA concentration and transcribe the mutant toxin sensor RNA *in vitro*.

PROTOCOL
Reagents and equipment needs are calculated per six student teams (appropriate excess included).

Equipment/glassware needed
1. Three sets of micropipettes 20-100 µl and 2-20 µl.
2. UV spectrometer.
3. Water bath set to 37 ºC.
4. Microcentrifuge.

Solutions needed
1. 1 ml ice-cold 75% ethanol.
2. 300 µl TE solution (10 mM Tris, 1 mM EDTA pH=8.0); RNase free.
3. Kit or polymerase and buffer for *in vitro* RNA synthesis.

<u>Precipitate plasmid DNA</u>
1. Spin plasmid DNA for 30 min at 4 °C at maximum speed. Carefully remove supernatant. This step pellets the DNA.
2. Add 50 µl ice-cold 75% ethanol.
3. Centrifuge plasmid DNA at 4 °C at maximum speed for 5 minutes.
4. Remove supernatant. This step removes residual salt.
5. Dry DNA pellet using a speed-vac (about 5 minutes) or on the bench.
6. Resuspend pellet in 20 µl TE buffer.

<u>UV spectroscopy</u> (this protocol is for Nanodrop spectrometer; for conventional UV spectrometer larger volumes of DNA need to be used)
1. Turn on spectrophotometer and choose nucleic acid assay setup.
2. Blank spectrophotometer with 2 µl millipore water.
3. Measure the absorbance of a water sample to insure the spectrometer is clean.
4. Use 2 µl plasmid DNA sample to measure the concentration of your plasmid DNA.
5. Record DNA concentration in g/l.

In vitro <u>transcription</u>
Protocol depends on kit used and will be provided in lab. See "transcription in practice" for a basic overview.

Note to the instructor
The experiment in Chapter 7 is designed to synthesize the ykkCD toxin sensor RNA *in vitro* using T7 RNA polymerase. The same protocol with minimal modifications can be used to synthesize any RNA sequence. The Ribomax T7 large scale RNA production system was purchased from Promega, but other vendors may be used. RNA production kits are preferable over using home-made T7 RNA polymerase, because they produce more RNA within a shorter period of time. To reduce cost, home-made T7 RNA polymerase may be used in place of an RNA production kit. Since RNA synthesis using a home-made T7 RNA polymerase takes about 1.5 hrs., we recommend starting the laboratory with DNA precipitation and giving pre-laboratory lecture while the RNA is synthesized. Alternatively, the long break may be used for an exam. Even if an RNA production kit is used this laboratory session has two breaks lasting 30 min each. Those breaks may be utilized to complete the restriction endonuclease worksheet.

Prelab Questions for RNA Transcription

Define the following terms.

1. RNA polymerase

/ 1 pts

2. Promoter

/ 1 pts

3. Transcription

/ 2 pts

4. Transcription terminator

/ 2 pts

5. Calculate how to mix a transcription reaction if the final reaction should have (in 20 ml).

2 ml T7 RNA polymerase
1 x transcription bufffer
1 mg linearized DNA

/ 4 pts

Assume that the transcription buffer stock is 2 x and the stock DNA concentration is 156 ng/ml. Complete volume (to 20ml) using water (if needed). Show your work.

RNA Synthesis
Lab Report Outline and Point Distribution

1. Several sentences defining the goal/purpose of this experiment. Make sure you indicate the role this step plays in the mini project (5 pts.).

2. Report the concentration of the linearized DNA. Calculate linearized DNA yield. Did you get a good yield in your assessment? Explain (8 pts.).

3. Based on the 260/280 absorbance ratio comment on the purity of the linearized DNA (5 pts.).

4. Your colleague had a satellite peak in the absorbance spectrum of his linearized DNA. Explain what might be the source of this contamination and how does this contamination bias his determination of DNA concentration (5 pts.). (Would he overestimate or underestimate the concentration due to the peak?)

5. Restriction endonuclease worksheet (27 pts.).

8 Purifying the ykkCD Mutant Toxin Sensor RNA and Evaluating its Purity Using Denaturing PAGE and UV spectrometry

8.1 Learning Objective

The purpose of this lab is to learn how to purify RNA samples and evaluate their purity using a denaturing polyacrylamide gel electrophoresis (urea PAGE). RNA molecules synthesized *in vitro* transcription are purified to remove the RNA polymerase, the DNA template and unused nucleotides (NTPs). The mutant toxin sensor RNA has to be purified for the following reasons: (1) The DNA template and unused nucleotides can interfere with accurate determination of RNA concentration using UV spectrometry, because these compounds also absorb the UV light at 260 nm. (2) The RNA polymerase, transcription buffer and unused NTPs may interfere with subsequent binding assays.

8.2 Mini Project Flowchart

The bolded block in the flowchart below highlights the role of the current experiment in the mini project.

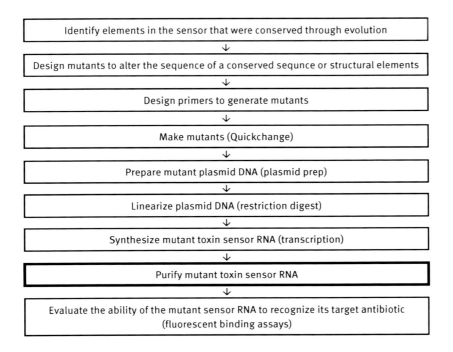

Identify elements in the sensor that were conserved through evolution
↓
Design mutants to alter the sequence of a conserved sequnce or structural elements
↓
Design primers to generate mutants
↓
Make mutants (Quickchange)
↓
Prepare mutant plasmid DNA (plasmid prep)
↓
Linearize plasmid DNA (restriction digest)
↓
Synthesize mutant toxin sensor RNA (transcription)
↓
Purify mutant toxin sensor RNA
↓
Evaluate the ability of the mutant sensor RNA to recognize its target antibiotic (fluorescent binding assays)

8.3 RNA Purification Methods

In vitro synthesized RNAs can be purified using various methods: denaturing polyacrylamide gel electrophoresis (denaturing PAGE), phenol/chloroform extraction or column purification. A brief description of each method is provided below.

8.4 Denaturing PAGE

Denaturing PAGE separates RNA molecules by their molecular weight (*Fig. 8.1A*). This method is used for both RNA purification and evaluation of RNA purity. In the latter case, RNAs are visualized using EtBr staining. Denaturing PAGE of RNAs is similar to agarose gel electrophoresis of DNAs. RNAs also migrate in the medium due to their negative charge from the negative to the positive electrode. There are two main differences: (1) the separation medium is polyacrylamide instead of agarose and (2) RNA samples are denatured so that the rate of migration in the gel is only affected by molecular weight and not by RNA structure. Since RNAs fold into stable, complex structures, to separate RNAs by molecular weight RNA molecules have to be denatured. Denaturation is achieved using urea - a chaotrop that competes with nucleobases for H-bonding thereby unfolding the RNA molecule (*Fig. 8.1B*). Polyacrylamide forms a similar net-like porous structure as agarose, but the pore size is significantly smaller accommodating separation of smaller nucleic acids.

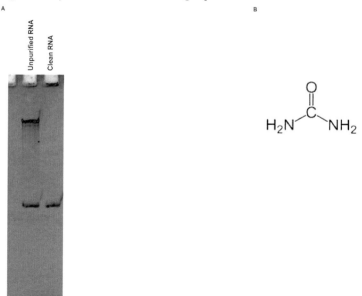

Figure 8.1: Denaturing polyacrylamide electrophoresis of RNAs. (A) RNA molecules are separated by molecular weight. The DNA template in the unpurified sample migrates on top, because of its larger molecular weight. (B) Urea is a chaotrop that unfolds RNAs by H-bonding with nucleobases.

Polyacrylamide gel is a polymer of acrylamide and methylenebisacrylamide; the methylenebisacrylamide is responsible for the crosslinks between acrylamide chains yielding a porous structure. The crosslinking ratio (ratio between acrylamide and methylenebisacrylamide) determines the pore size of the medium. A higher crosslinking ratio denotes a medium with a smaller pore size. That is designed to separate smaller RNAs (*Fig. 8.2*).

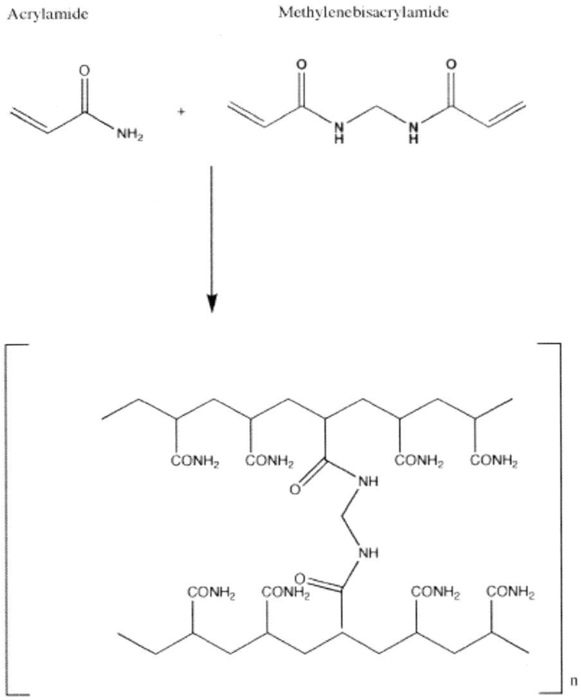

Figure 8.2: Polyacrylamide gel electrophoresis. The polyacrylamide medium is generated by free-radical polymerization. The ratio of acrylamide: methylenebisacrylamide determines the pore size of the gel.

Denaturing PAGE produces the cleanest RNA. The transcription mixture is separated on a denaturing PAGE. The desired RNA band (toxin sensor RNA) is excised (cut) from the gel and the RNA is eluted. This method removes the RNA polymerase, the DNA template, unused nucleotides and any *in vitro* synthesized RNAs that differ in size from the toxin sensor RNA (mostly RNAs that were the product of degradation). Denaturing PAGE purification of RNA is time consuming and only used when high-purity RNA is an absolute requirement for structure determination, time-resolved fluorescent spectroscopy etc.

8.5 Phenol/chloroform Extraction

Phenol/chloroform extraction removes the polymerase and most unused nucleotides from the reaction mixture. Since some unused nucleotides and the DNA template is still in the mixture, accurate determination of RNA concentration is not possible using this purification method (recall, all of these nucleic acids absorb the UV light at 260 nm). To evaluate the ability of the mutant toxin sensor RNA to recognize the antibiotic tetracycline, the binding affinity between the mutant sensor and tetracycline has to be determined accurately. Since binding affinity measurements require precise knowledge of RNA concentration, phenol/chloroform extraction does not produce adequate RNA quality for our purposes.

8.6 Column Purification

Our method of choice for RNA purification is column purification. This method is very similar to purification of plasmid DNAs during plasmid preparation. When RNA is synthesized *in vitro* for biochemical purposes, the goal is to get high concentrations of the target RNA in water. In contrast the RNA after synthesis is contaminated with the polymerase, the DNA template, unused nucleotides and a buffer. During column purification all of these contaminants are removed. Column purification of RNAs utilize the fact that *in vitro* synthesized RNAs are highly negatively charged (compared to proteins) and fairly small (compared to the template DNA). Column purification is advantageous; because it is quick, it removes the DNA template and unused nucleotides, thus enables accurate determination of RNA concentration.

Make sure to follow the protocol exactly as written. Always wear gloves and store RNA on ice to prevent degradation of your sample.

The exact protocol of RNA purification is dependent on the kit used and will be provided in the lab. A general outline of procedure is presented below.

1. Column purification of RNA usually starts with binding the transcription mixture to an anion exchange column. This separation method based on the difference in charge between protein, nucleotides and nucleic acids (polymers of nucleotides).
2. The template DNA and the RNA binds to the column, while unincorporated nucleotides (too small) and the polymerase (not charged enough) flow through.
3. The DNA template is much larger (more negatively charged) than the synthetic RNA and this means it is not eluted from the column.

What are we going to do?

Today we will purify the *in vitro* synthesized toxin sensor RNA via column purification then evaluate RNA purity using denaturing urea PAGE and determine RNA concentration using UV spectrometry.

PROTOCOLS

Reagents and equipment needs are calculated per six student teams. There is ~20% excess included.

Equipment/glassware needed

1. Six disposable test tubes at least 10 ml volume
2. Three sets of micropipettes 20-100 μl and 2-20 μl
3. Three sets of mini protean 3 PAGE apparatus
4. Power supply (to run urea PAGE)
5. UV spectrometer
6. Gel documentation system
7. Twenty 1.5 ml centrifuge tubes (Eppendorf)
8. Heating block (to boil samples)

Solutions needed

1. 3 l 1 x TBE (Tris-borate-EDTA) pH=8.0
2. 30 μl 1% EtBr
3. 300 μl urea gel loading dye (2x); (6 M urea, 0.5 x TBE, 1% bromophenol blue)
4. Urea gel solution mix; (6 M urea, 0.5 x TBE, 10% acrylamide: bis acrylamide 19:1 crosslinking ratio)
5. 1 ml 10% ammonium persulfate (APS) freshly made
6. 50 μl TEMED

Column purification of RNA

Protocol depends on the kit used and will be provided in the lab. See the top of the page for a basic overview.

Determine RNA concentration using UV spectroscopy

RNA concentration is determined in a similar manner as DNA concentration. RNAs also absorb at 260 nm due to the heterocyclic aromatic nucleobases. To calculate RNA concentration the Beer-Lambert law is used: $A(@260 \text{ nm}) = \varepsilon*c*l$ where $l=1$ cm, c is the RNA concentration in mole/liter and $\varepsilon = 1149800$ for the ykkCD toxin sensor RNA.

Denaturing PAGE of RNA

Make the gel

1. Assemble gel plates and check them for leaking with EtOH.
2. Pipette 10 ml of urea gel mixture into a test tube.

3. Add 100 µl 10% APS and 10 µl TEMED to initiate polymerization.
4. Mix well and pour in between glass plates. Place in comb. WATCH OUT, THIS IS QUICK! WAIT UNTIL GEL POLYMERIZES!
5. Once gel is polymerized assemble gel gasket.
6. Pour 0.5 x TBE running buffer into gasket.
7. Wash wells.
8. Pre-run gel for 30 min at 15 W. This step gets rid of any acrylamide that did not form a polymer and any free radical produced during polymerization.

Prepare sample and run the gel. Run 10 µl unpurified and 10 µl purified RNA on the gel.
1. Mix 10 µl RNA with equal volume of urea dye (6M urea, 0.5 x TBE, 1% bromophenol blue).
2. Boil samples for 5 min in the heating block. Quick spin samples.
3. Load samples onto the gel. REMEMBER WHERE YOU LOADED YOUR SAMPLE!
4. Run gel for 15 min at 15 W or until the dye reaches the bottom of the gel.
5. Place the gel into 50 ml staining solution (50 ml TB buffer and 10 µl EtBr) for 10 min.
6. Scan gel.

Precautions
1. *Acrylamide is a neurotoxin*
2. *EtBr is a carcinogen/mutagen*
MAKE SURE TO WEAR GLOVES

Note to the instructor
The experiment in Chapter 8 is designed to purify the ykkCD toxin sensor RNA, but the protocol with minor modifications could be used for any other RNA molecule. Due to the time requirements of making and pre-running a denaturing PAGE it is recommended to start the laboratory session with making the gel then purifying the RNA while the gel pre-runs. Qiagen nucleotide removal kit and GeneJet RNA Cleanup kit (Thermo Scientific) were used by the authors, but similar kits from alternate vendors may also be used. It is preferable that each student team make and run a denaturing PAGE to increase the educational value of the laboratory, but if supplies are limited six student teams may share a gel. The denaturing PAGE may also be made by teaching assistant or the instructor ahead of time. In this case however the laboratory is more appropriate to be called a demonstration of RNA purification.

Prelab Questions for RNA Purification

1. What is "urea gel electrophoresis (denaturing PAGE)"?

/ 2 pts

2. What do you remove from the RNA sample during RNA purification? Name two things.

/ 2 pts

3. What type of precautions do you have to observe when handling acrylamide? Explain your answer.

/ 3 pts

4. To run a denaturing PAGE you need 1 L 0.5 x TBE buffer. The stock solution is 10 x TBE. Explain how you make your buffer (show your calculation).

/ 3 pts

RNA Purification
Lab Report Outline and Point Distribution

1. Several sentences defining the goal/purpose of each procedure. Make sure you clearly indicate the role of this step in the mini project (5 pts.).

2. Briefly describe the methods available for RNA purification. Indicate why we choose column purification in the lab (7 pts.).

3. Draw a flowchart of the RNA purification. Make sure to indicate the name of each buffer used and the volume of buffer needed (10 pts.).

4. Include a picture of your denaturing PAGE gel scan with your samples marked (4 pts.).

5. Evaluate the purity of your mutant ykkCD sensor RNA by comparing the bands before and after purification. Explain your answer (5 pts.).

6. Report the concentration of your purified ykkCD RNA. Make sure you use the Beer-Lambert law and the molar extinction coefficient provided to carry out your calculation. Show your work (5 pts.).

7. Calculate ykkCD sensor RNA yield. Show your work (5 pts.).

8. Report the 260/280 absorbance-ratio of the purified ykkCD RNA. Evaluate RNA purity based on this value. What types of contaminants could cause a peak around 180 wavelength (5 pts.)?

9. Troubleshooting question: Your colleague performed RNA purification with the same kit you used. He later performed a binding assay and determined that his ykkCD RNA (wild type) had low affinity for tetracycline. His denaturing PAGE and 260/280 absorbance values each indicated clean RNA and he appeared to have good yield. He later determined that his RNA solution contained ethanol. In your judgment, which step of the purification did he make the mistake? Is there any way to rescue his RNA (4 pts.)?

9 Evaluating the Ability of the ykkCD Toxin Sensor to Recognize the Antibiotic Tetracycline Using Fluorescent Quenching

9.1 Learning Objective

The goal of the mini project was to identify elements in the toxin sensor that are essential to recognize the antibiotic tetracycline. You identified elements in the sensor that did not change throuhout evolution (invariable blocks). You subjected them to the site-directed mutagenesis, modified the sequence of the toxin sensor DNA and made the mutated toxin sensor RNAs. During this lab you will evaluate how well this mutant sensor is able to recognize the antibiotic tetracycline. If the mutant sensor is still able to recognize tetracycline, it means that the mutated invariable block was not essential for sensory function. If however, the mutated sensor fails to recognize tetracycline, it means that the mutated invariable block was essential for sensory function. You will perform binding affinity assays using fluorescence quenching to evaluate the affinity of the mutant sensor toward tetracycline.

9.2 Mini Project Flowchart

The bolded block in the flowchart below highlights the role of the current experiment in the mini project.

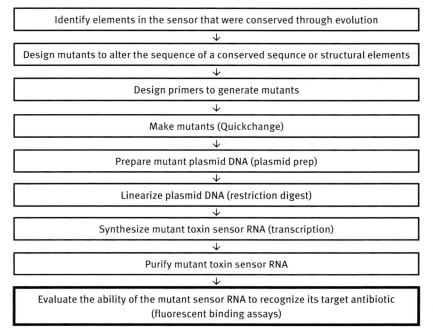

9.3 What is Binding Affinity (K$_D$)?

To estimate how well the mutated sensor RNA recognizes the antibiotic tetracycline, you will measure the binding affinity (K$_D$) of the tetracycline-sensor RNA complex. Binding affinity is the equilibrium dissociation constant. There is an inverse relationship between the strength of the interaction and the numeric value of the binding affinity: if the binding affinity is a small number, it means that addition of a small amount of RNA to tetracycline results in a high concentration of the sensor RNA-tetracycline complex. In other words, it only takes a small amount of tetracycline to form enough complex with the sensor RNA to trigger production of the efflux pump that in turn gets rid of tetracycline. *Therefore, a small K$_D$ value reflects a strong interaction between the antibiotic tetracycline and the sensor RNA. Likewise, a large K$_D$ value reflects a week interaction between tetracycline and the sensor RNA.* In other words if the K$_D$ value is large, it takes a large amount of tetracycline to form enough tetracycline-RNA complex to trigger bacterial defense that gets rid of tetracycline. To measure binding affinity, one of the two reactants (tetracycline in our case) is kept at a constant low concentration while the concentration of the other reactant (sensor RNA) is varied. To determine the binding affinity, the fraction of tetracycline that is in complex with the sensor RNA is plotted against the RNA concentration. The dissociation constant is the sensor RNA concentration that forces 50% of tetracycline to form complex with the sensor RNA (*Fig. 9*).

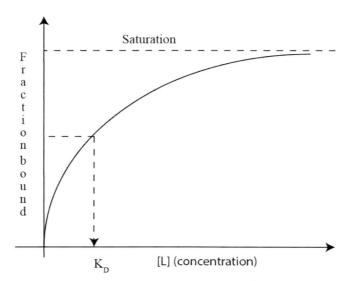

Figure 9.1: The K$_D$ value is the sensor RNA concentration that forces 50% of the tetracycline to form the tetracycline-ykkCD RNA complex.

9.4 What is Fluorescence?

Fluorescence is a natural phenomenon. Some compounds, usually heterocyclic aromatic molecules, are able to absorb light of a specific energy and subsequently emit light that is *lower* in energy (larger wavelength) than the light absorbed. The wavelength of the light absorbed by the fluorescent compound is referred to as the *excitation wavelength*. The wavelength of the light emited by the fluorescent compound is referred to as the *emission wavelength* (*Fig. 9.2*).

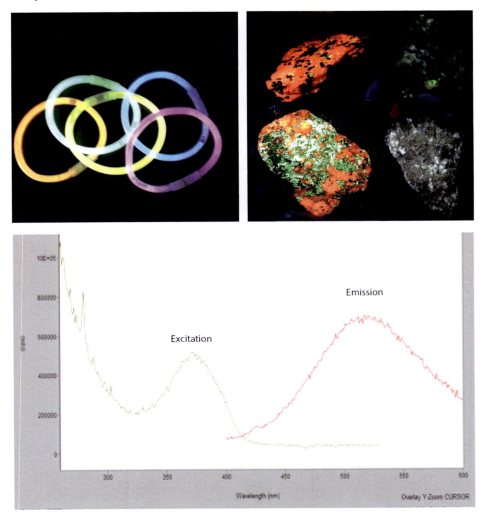

Figure 9.2: Fluorescence is a natural phenomenon (top). Fluorescent compounds emit light that is lower in energy, and of higher wavelength, than the light absorbed (bottom).

During fluorescence, the incoming photons (excitation wave) excite electrons in the molecule to the higher energy excited states. The electrons lose some energy via nonradiative decay to reach the lowest energy excited state. From here, the excited electrons emit energy in form of photons (emission wave) to return to the ground state. The energy of the emission wave is lower than that of the excitation wave, because some energy was lost due to nonradiative decay, usually heat (*Fig 9.3*).

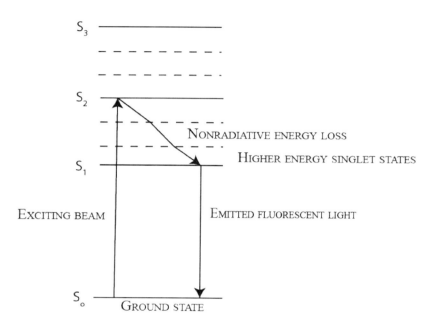

Figure 9.3: Jablonski diagram illustrates the electronic states of a molecule undergoing fluorescence. The electron is excited from the ground state to a higher energy excitation state. From there it releases energy through nonradiative decay then emits light (fluorescent light) to return to the ground state.

9.5 How Do We Measure Binding Affinity of the Tetracycline-Sensor RNA Complex?

To measure the binding affinity of the tetracycline-sensor RNA complex, you need to determine the fraction of tetracycline that is in complex with the toxin sensor RNA (fraction bound). To measure fraction bound you will take advantage of the natural fluorescence of tetracycline: once the sensor RNA forms a complex with tetracycline, the fluorescence of tetracycline decreases (*quenching*). The amount of quenching is proportional to the fraction of tetracycline that is bound to the sensor RNA (*Fig 9.4*).

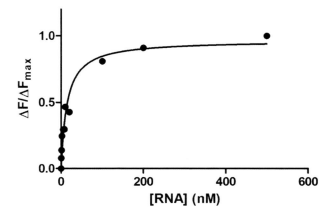

Figure 9.4: Diagram of fluorescent binding assay. Once tetracycline is bound to the sensor its natural fluorescence decreases leading to fluorescent quenching.

To determine the dissociation constant (K_D) you need to plot the amount of quenching (proportional to the fraction of tetracycline bound to the sensor RNA) against the sensor RNA concentration. This plot should shape as a hyperbola (saturation curve). Then you should fit the Equation 1 to determine the K_D value (*Fig. 9.5*).

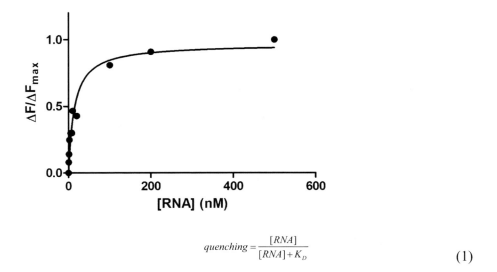

$$quenching = \frac{[RNA]}{[RNA] + K_D} \tag{1}$$

Figure 9.5: Determination of K_D value. The amount of quenching is plotted against the sensor RNA concentration. The K_D value is the sensor RNA concentration where 50% of quenching is achieved.

9.6 How do We Evaluate Binding Affinity?

The goal of the mini-project was to identify parts of the toxin sensor that are essential for recognizing the antibiotic tetracycline. To reach the goal, you modified (mutated) evolutionary conserved parts of the toxin sensor and evaluated how well these mutated sensors were able to retain their ability to recognize tetracycline by measuring the K_D of the mutant sensor RNA-tetracycline complex. If the mutated sensor retain its ability to recognize tetracycline (small K_D value), the part of the sensor targeted for mutagenesis was not essential to recognize tetracycline. Likewise, if the mutated sensor lose its ability to recognize tetracycline (large K_D value), the part of the sensor targeted for mutagenesis was important for tetracycline recognition (*Fig 9.6*).

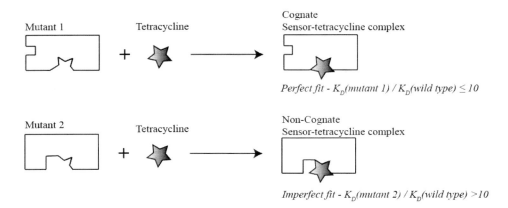

Perfect fit - K_D(mutant 1) / K_D(wild type) ≤ 10

Imperfect fit - K_D(mutant 2) / K_D(wild type) > 10

Figure 9.6: Schematics illustrating how to interpret the results of the binding affinity assays: Mutant 1 binds strongly to tetracycline, thus we conclude that the nucleotides changed were not essential for tetracycline recognition. In contrast, mutant 2 did not bind strongly to tetracycline. This means the nucleotides changed were probably important for tetracycline recognition.

9.7 How do We Analyze Data?

Based on the data shown on *Fig. 9.7*, the mutant sensor-tetracycline complex has a K_D value of about a 100 nM. Compared to a binding affinity (10 nM) reported for the wild-type sensor RNA tetracycline complex, the mutant sensor slightly lost its ability to recognize the antibiotic tetracycline. Based on these data, we conclude that the part of the toxin sensor mutated is likely to be important for tetracycline recognition. To determine the K_D value of the mutant sensor-tetracycline complex more accurately, a higher mutant sensor RNA concentrations should be used in the future.

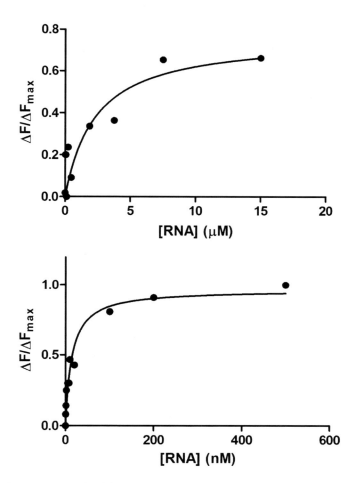

Figure 9.7: Example of data analysis. The binding affinity of the mutant sensor RNA-tetracycline complex is elevated (upper panel) compared to that of the wild-type sensor RNA tetracycline complex (lower panel). Elevated K_D value corresponds to weakened binding to tetracycline. Thus we conclude that the mutated nucleotides were important for tetracycline recognition.

What are we doing today?
1. Set up binding assays
2. Use demo data to fit binding curve and determine K_D

PROTOCOLS
Reagents and equipment needs are calculated per six student teams. There is ~20% excess included.

Equipment/glassware needed
1. Micropipettes 20-100 µl and 2-20 µl
2. Fluorescent plate reader
3. 96-well plates

Solutions needed
1. 5 x reaction buffer (100 mM Tris, pH=8.0; 500 mM KCl; 5 mM $MgCl_2$)
2. 20 nM tetracycline (in DMSO)
3. Ribolock RNase inhibitor (Fermentas, 40 U/µl)

Binding assay setup
Each student should set up 2 assays per mutant and one assay using the wild type ykkCD RNA sensor.

1. Prepare 2 solutions for each row (assay)
    RNA containing solution
    20 µl 5x reaction buffer
    10 µl 20 nM tetracycline
    Up to 70 µl or 1 µM final concentration sensor RNA
    0.5 µl Ribonuclease inhibitor
    Water to 100 µl if needed

    No RNA solution
    120 µl 5x reaction buffer
    60 µl 20 nM tetracycline
    3 µl Ribonuclease inhibitor
    417 µl water

2. In each row perform serial dilution using the solutions above as follows

Well	1	2	3	4	5	6	7	8	9	10	11	12
RNA containing solution	100 µl	50 µl of well 1	50 µl of well 2	50 µl of well 3	50 µl of well 4	50 µl of well 5	50 µl of well 6	50 µl of well 7	50 µl of well 8	50 µl of well 9	50 µl of well 10	NO
No RNA solution	0 µl	50 µl	50 µl	50 µl	50 µl	50 µl	50 µl	50 µl	50 µl	50 µl	50 µl	50 µl

CALCULATE RNA CONCENTRATION OF EACH WELL ACCURATELY

4. Seal tray, cover with aluminum foil and incubate for 72 hrs. at 4 ºC.

Note to the instructor
The experiment in Chapter 9 was designed to determine the binding affinity of the mutant ykkCD sensor-tetracycline complex. The same protocol with minimal modifications can be used to measure binding affinity of any other fluorophore-macromolecule complex using fluorescent quenching. The binding assay worksheet may be incorporated into a lecture course that teaches binding affinity. 96-well plates used were manufactured by Corning Inc. (3991), but as long as the plates are black, flat bottom and have 96-wells, any other vendor might be used. Usage of RNase inhibitor is important to prevent RNA degradation especially considering the limited research experience of the experimenters. Assays may be read out by teaching assistants or by students that take the laboratory course at a different time of the week. A Tecan Infinite F200 plate reader was used to read out the assays; excitation wavelength was 380 nm, emission wavelength was 535 nm.

Analysis of Binding Experiments

In a common binding experiment to a set amount of one reagent (A) successive amounts of a second reagent (B) are added; the fractional amount bound (AB) is then measured. Since the second reagent (B) is added until all the first reagent (A) is bound, A is said to be "saturated" with B. Often the goal of the experiment is to determine the dissociation equilibrium constant, K_D.

Whether we study a binding protein or RNA, etc., the binding equilibrium can be treated in the same way:

$$\text{fraction of A that is bound} = \frac{[B]}{K_D + [B]} \tag{2}$$

Equation 2 predicts a hyperbolic binding curve (see below). This is sometimes termed a "saturation" curve because A is finally saturated with B.

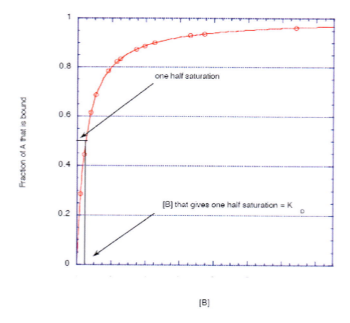

Figure 9.8: Saturation curve depicting binding of a ligand to a receptor. K_D value indicates the ligand concentration at which the binding site on a particular protein is half occupied.

The dissociation equilibrium constant is equal to the concentration of B that gives one half-saturation. For the RNA sensor binding assay, you will use a set amount of tetracycline to which increasing amounts of RNA will be added. The fluorescence of tetracycline will decrease as it is bound by the RNA.

Fluorescent quenching is used to measure the fraction of tetracycline bound by RNA.

$$Quenching = fraction\ of\ tetracycline\ bound\ by\ RNA = \frac{\Delta\ Fluorescence}{\Delta\ Fluorescence_{max}} \tag{3}$$

where

$$\Delta\ Fluorescence = fluorescence\ for\ no\ RNA\ \text{-}\ fluorescence\ for\ a\ specific\ [RNA] \tag{4}$$

and

$$\Delta\ Fluorescence_{max} = fluorescence\ with\ no\ RNA\ \text{-}\ fluorescence\ for\ largest\ [RNA] \tag{5}$$

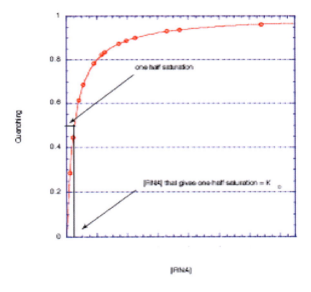

Figure 9.9: Saturation curve depicting binding of tetracycline to the sensor RNA. K_D value indicates the ligand concentration where 50% of the sensor RNA is bound with tetracycline.

The RNA-Tetracycline Binding Worksheet contains problems that allow you to work through these calculations.

Once you have plotted the hyperbolic binding curve, you need to use a double reciprocal plot; this generally gives a better measure of K_D. The reciprocal of the binding equation predicts a straight line plot:

$$\frac{1}{Quenching} = \frac{K_D + [RNA]}{[RNA]} = K_D \frac{1}{[RNA]} + 1 \tag{6}$$

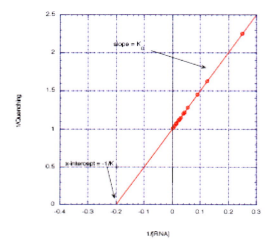

Figure 9.10: Linear representation of a saturation curve. The K_D value may be determined in two different ways: (1) the slope of the graph equals the K_D value or (b) the x-intercept is -1/K_D.

Binding Assays Prelab

Work with the following data from binding assays using two sensor RNA mutants. Please use a computer graphing program. Mutations are highlighted on the figure below.

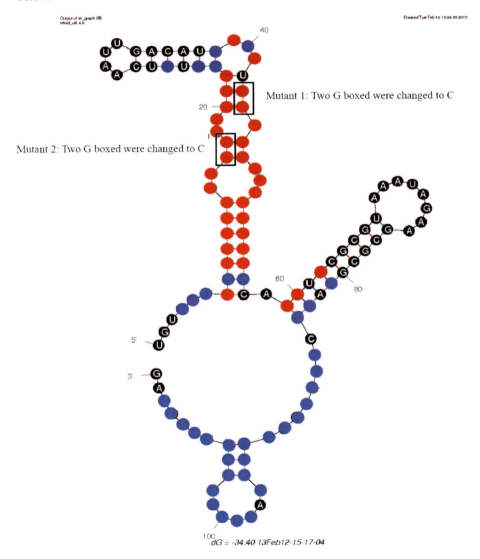

Mutant 1: Two G boxed were changed to C

Mutant 2: Two G boxed were changed to C

Figure 9.11: *B. subtilis* **ykkCD sensor RNA secondary structure:** Color coding is as follows: red, 100% sequence conservation; blue, >80% sequence conservation and black, no significant sequence conservation.

1. (10 pts.) The fluorescence from tetracycline was measured as a function of [RNA] for Mutant #1 and the following results were obtained (Quenching = $\Delta Fl/\Delta Flmax$):

[RNA] (nM)	Fluorescence	ΔFluorescence	$\Delta Fl/\Delta Fl_{max}$	1/[RNA]	$\Delta Fl_{max}/\Delta Fl$
0.0000	1541.0				
10.000	1473.0				
20.000	1497.0				
50.000	1446.0				
80.000	1443.0				
100.00	1368.0				
200.00	1406.0				
500.00	1304.0				
700.00	1350.0				
900.00	1300.0				
1100.0	1295.0				

a) Plot quenching versus [RNA] and estimate the K_D for tetracycline-RNA binding. Show how you estimated the dissociation constant.

b) Use a double reciprocal plot to get a better estimate of K_D.

2. (15 pts.) Mutant #2 was also studied. The stock [RNA] was 2.4×10^{-5} M. The procedure for this experiment matched the one you are running in lab. Prepare a double reciprocal plot and determine the K_D for mutant #2.

RNA microliters into total	[RNA] (nM)	Fluorescence	ΔFluorescence	ΔFl/ ΔFlmax	1/[RNA]	ΔFlmax/ ΔFl
Well 1: 50µL into 100µL		1570.0				
Well 2: 50µL Well1 into 100µL		1568.0				
Well 3: 50µL Well 2 into 100µL		1570.0				
Well4: 50µL Well 3 into 100µL		1600.0				
Well 5: 50µL Well 4 into 100µL		1610.0				
Well 6: 50µL Well 5 into 100µL		1690.0				
Well7: 50µL Well 6 into 100µL		1720.0				
Well 8: 50µL Well 7 into 100µL		1752.0				
Well 9: 50µL Well 8 into 100µL		1830.0				
Well 10: 50µL Well 9 into 100µL		1864.0				
Well 11: no RNA		1875.0				

3. (5 pts.) Which mutation has a larger impact on tetracycline binding? Briefly explain.

YkkCD sensor RNA - Tetracycline Binding
Lab Report Outline and Point Distribution

1. Several sentences defining the goal/purpose of each procedure. Describe the role this step plays in the mini project (5 pts.).

2. Describe fluorescence (about five sentences; 5 pts.).

3. Show an example of your calculations for RNA concentration in the binding assay (3 pts.).

4. Use a computer to plot (20 pts. total):
a. Quenching vs. [RNA] for each assay; determine the K_D from the curves. Report error and standard deviation!

b. $\dfrac{1}{\text{Quenching}}$ vs. $\dfrac{1}{\text{[RNA]}}$ for each assay; determine the K_D from these double reciprocal plots. Report error and standard deviation!

c. Do you consider your K_D measurement reliable? Briefly explain. How would you modify your experiment to more accurately determine K_D? For example, could you change the [RNA] range? Briefly explain (10 pts.).

5. In your judgment how did the mutation affect the ability of the sensor to recognize the antibiotic tetracycline? Do you think the mutated region was important for antibiotic recognition? How does the K_D of the mutant sensor – tetracycline complex relate to that of the wild type sensor - tetracycline complex? (7 pts.).

10 Evaluating Antibiotic Binding to Blood Serum Albumin Using Fluorescence Spectroscopy

10.1 Learning Objectives

In this laboratory, you will study one of the most important functions of proteins. Proteins bind specific small molecules in a very selective fashion. This laboratory focuses on the major extracellular protein in the blood stream, human serum albumin. To study binding, we have chosen a sensitive optical measurement, fluorescence. You will use the wealth of data from this sensitive technique to study the details of antibiotic binding to albumin.

10.2 Biological Role of Serum Albumin

Albumin is the major circulating protein in the blood stream comprising about half of the total serum protein. Its structure contains many hydrophobic pockets which bind a variety of biological molecules. Several examples follow:
- By binding to albumin, fatty acids can be transported at high concentrations (free fatty acids are soluble to about 10^{-6} M while the fatty acid - albumin complex is soluble into the millimolar range).
- When red cells die and lyse, excess of the heme is released. Albumin binds this excess removing it from the blood stream.
- Hydrophobic hormones, such as the thyroxines or steroids are often bound by albumin.
- Many anionic, hydrophobic drugs bind to albumin. This binding has a major impact on the drugs' effectiveness. This binding can slow drug distribution to the tissues, reduce drug clearance and cause an overall loss of drug efficacy. Interactions between drugs are also mediated by serum albumin. For example, one drug may cause an increase in availability of a second drug (thus, increasing the effective drug dose) if the drugs compete for the same binding site of serum albumin.

During this laboratory you will masure binding affinity between albumin and a common antibiotic, levofloxacin (Levaquin®).

10.3 Fluoroquinoline Antibiotics

Levofloxacin (Levaquin®) is a member of the fluoroquinoline class of antibiotics. These drugs have been known for about 50 years, but especially effective derivatives were not widely available until the 1980s. These are "second generation" fluoroquinolines and include the common drugs, ciprofloxacin (Cipro®) and ofloxacin (Floxin®).

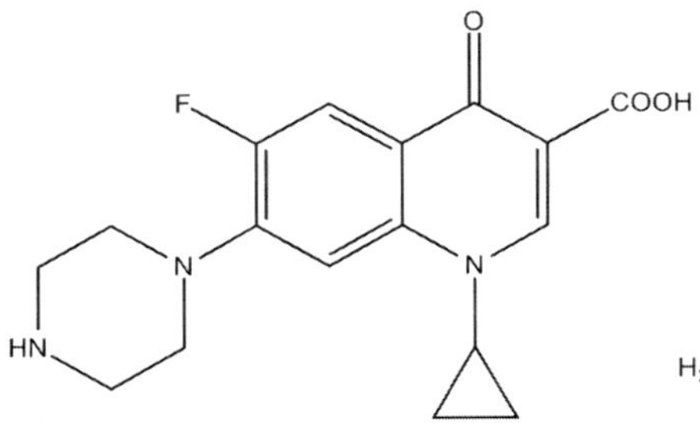

Ciproflaxin

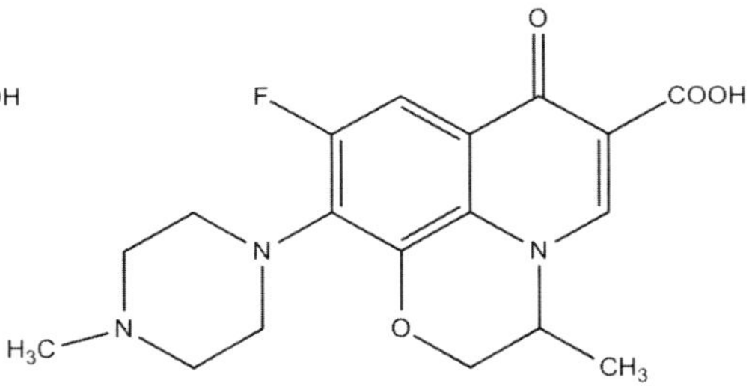

Ofloxacin

Levofloxacin (Levaquin®) is a "third generation" fluoroquinolone and is simply the biologically-active isomer of ofloxacin (a racemic mixture).

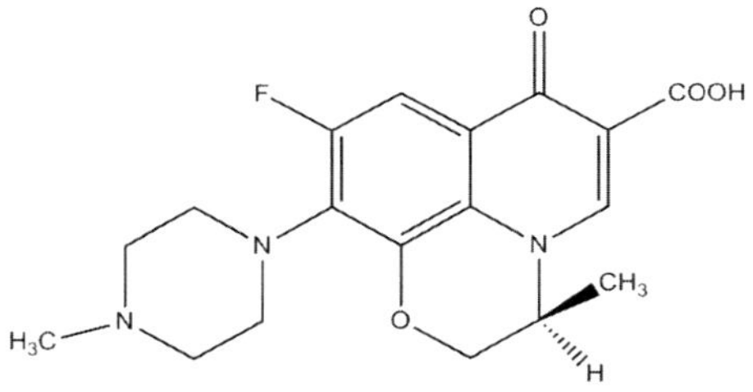

Levafloxacin

The fluoroquinolones have a unique site of action, inhibiting the bacterial DNA gyrase (topoisomerase type II) and topoisomerase IV. These enzymes unwind DNA and are required for DNA replication. The fluoroquinolines block DNA unwinding and, thus, block bacterial replication. They are most effective against Gram-negative bacteria although the later generation drugs also are effective against some Gram-positive and anaerobic bacteria.

10.4 Protein Structure, Aromatic Amino Acids, and Fluorescence

Proteins are polymers of amino acids. The amino acid side chains (R groups) are primarily responsible for unique properties of each protein. For example, if a protein has many amino acids with alkyl side chains, then the protein is relatively hydrophobic. Or, if the protein has many amino acids with carboxylate-containing side chains, then the protein tends to be negatively charged.

The aromatic amino acids, phenylalanine, tyrosine and tryptophan, contribute to the unique property of proteins, fluorescence. Fluorescence is a process by which a molecule absorbs light and then emits the light again (typically at a longer wavelength; *Fig. 10.1*).

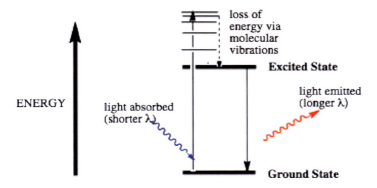

Figure 10.1: An energy diagram showing the energy transfer during a fluorescent process. Absorbed light excites an electron to a higher energy orbital. Some of the absorbed energy is lost to molecular vibrations. The remainder of the absorbed energy is released as light when the excited electron returns to the ground state.

The aromatic amino acids have an absorbance peak between 260 nm to 290 nm. They emit light maximally from about 290 nm to about 350 nm.

10.5 Measuring Fluorescence

To measure fluorescence, you always set <u>two</u> wavelengths, the wavelength of light irradiating the sample (the excitation or absorbance λ) and the wavelength of emitted light to be measured (emission λ). To scan an emission spectrum, the absorbance/ excitation λ is fixed and a monochromator scans through the emission wavelengths (*Fig. 10.2*).

To scan an excitation spectrum, the emission λ is fixed and a monochromator scans through wavelengths of light that irradiate the sample (*Fig. 10.3*).

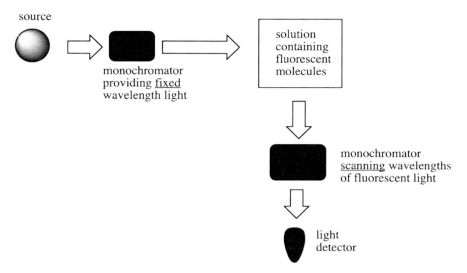

Figure 10.2: Block diagram showing the major spectrofluorometer components used to measure an emission spectrum. Fluorescent light is emitted from all sides of the solution containing fluorescent molecules and is measured at right angles to the excitation light to minimize background interference.

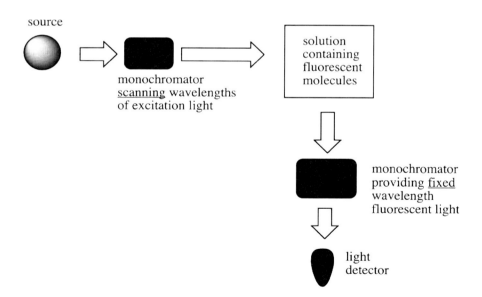

Figure 10.3: A block diagram showing the major spectrofluorometer components used to measure an excitation spectrum. Note that a typical research-grade spectrofluorometer has <u>two</u> monochromators, one for excitation light and one for emission light.

Fluorescent molecules, like the aromatic amino acids, have characteristic emission and excitation spectra - these spectra can serve to identify molecules. For example, the shapes of the peaks as well as the peak wavelengths are characteristic of specific molecules. Tryptophan has a <u>broad</u> emission peak at about 350 nm while tyrosine has a <u>narrow</u> emission peak at about 305 nm. The wavelength difference between excitation and emission also is characteristic. Depending on a molecule's electronic energy levels, the difference between excitation and emission may be large (much energy is lost before light is emitted) or small (little energy is lost before light is emitted). For example, tryptophan has about a 60 nm difference between excitation and emission wavelength while tyrosine has only about a 15 nm difference (*Fig. 10.4*). Tryptophan loses more energy before emission while tyrosine loses less.

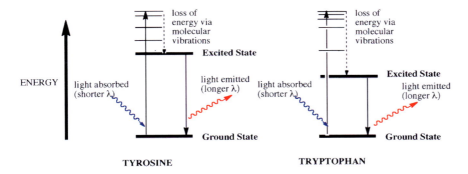

Figure 10.4: Energy diagrams that qualitatively compare the tyrosine fluorescence with the tryptophan fluorescence. The tyrosine fluorescence (shown on the left) is higher in energy (shorter wavelength) than the tryptophan fluorescence (shown on the right). Once excited, tryptophan loses more energy to vibrations than does tyrosine.

10.6 Synchronous Spectroscopy

Synchronous spectroscopy selects for fluorophores with a specified energy difference between excitation and emission. This technique depends on maintaining a constant difference between the excitation wavelength and the emission wavelength. The excitation monochromator scans wavelengths simultaneously and at the same speed as the emission monochromator. Since the monochromators are synchronized with each other (synchronous spectroscopy) a constant $\Delta\lambda$ is maintained. Using a $\Delta\lambda = 15$ nm means the spectrofluorometer is sensitive to groups that fluoresce light after a small shift in wavelength, i.e., tyrosine. Setting a $\Delta\lambda = 60$ nm allows the spectrofluorometer to "see" groups that fluoresce light after a large shift in wavelength, i.e., tryptophan. By setting a given $\Delta\lambda$, the spectrofluorometer is able to monitor select, specific amino acids.

10.7 Data Analysis

The goal of this laboratory is to monitor binding of levofloxacin to albumin.

$$\text{Levofloxacin} + \text{Albumin} \rightleftharpoons \text{Levofloxacin - Albumin}$$

Levofloxacin binding is proportional to the change in fluorescence. Since the fluorescence decreases, the change is called fluorescence quenching. To measure binding of levofloxacin by albumin as a function of levofloxacin concentration, we need to plot fluorescence quenching versus the concentration of the levofloxacin in the cuvette. In general, binding between a protein (P) and a ligand (L) fits the following equation:

$$P + L \rightleftharpoons PL$$

where

$$K_D = \frac{[P][L]}{[PL]} \tag{1}$$

In most experiments, [L] (the *independent* variable) is varied and [PL], the *dependent* variable, is measured. Equation 1 is algebraically rearranged to put the dependent variable on the left and the independent variable on the right. Note that Equation 2 has the same form as the Michaelis-Menten equation (*Fig. 10.5*).

$$[PL] = \frac{[PL_{max}][L]}{K_D + [L]} \tag{2}$$

and

$$\frac{[PL]}{[PL_{max}]} = \frac{[L]}{K_D + [L]} \tag{3}$$

It is important to recognize that the dissociation constant, K_D, is numerically equal to the [L] that yields $[PL] = \frac{1}{[PL_{max}]}$. So, the K_D can be determined from both graphs. Since fluorescence quenching $\frac{\Delta F}{[\Delta F_{initial}]}$ is equal to $\frac{[PL]}{[PL_{max}]}$,

$$\frac{Fl_{initial} - Fl}{Fl_{initial}} = \frac{\Delta Fl}{\Delta Fl_{initial}} = \frac{[levofloxacin]}{K_D + [levofloxacin]} \tag{4}$$

This equation is analogous to Equation 3 and leads to a graph as shown in *Fig. 10.6*.

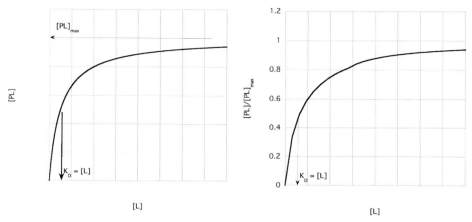

Figure 10.5: Common graphical representations of ligand binding to a protein. In both graphs the dependent variable is a function of [PL] while the independent variable is [L]. The graph on the right plots the ratio $\frac{[PL]}{[PL_{max}]}$. When plotted this way, the Y-axis varies from 0 to 1.

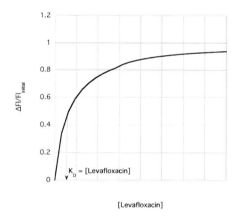

Figure 10.6: Graphical representation of levofloxacin binding to human serum albumin: The independent variable is the levofloxacin concentration. The dependent variable is fluorescence quenching which is equal to $\frac{[PL]}{[PL_{max}]}$.

This plot is similar to the two plots you created for your titrations. However, it is <u>not</u> the same. This plot graphs the *free* [levafloxacin], that is the total [levafloxacin] in the cuvette minus the [levafloxacin] bound to albumin. Your two plots graphed the *total* [levafloxacin] in the cuvette. Your concentrations must be corrected before an equilibrium constant can be calculated. Following rearrangement and substitution of Equation 4, Equation 5 is derived:

$$\frac{Fl_{initial}}{Fl_{initial}} = \frac{\Delta Fl}{\Delta Fl_{initial}} = \frac{[Levofloxacin]}{K_D + [Levofloxacin]} \tag{5}$$

$$\frac{Fl_{initial}}{Fl} = \frac{1}{K_D} [Levofloxacin]_{curvette} \frac{Fl_{initial}}{\Delta Fl} - \frac{1}{K_D} [Albumin]_{total}$$ (6)

Equation 6 allows calculation of K_D. Furthermore, this equation is set in a linear form using the slope-intercept equation, $y = m\,\mathbf{x} + b$. Note that this equation uses $Fl_{initial}$, ΔFl and Fl. Fluorescence spectroscopy can provide information concerning protein conformation changes. Both tyrosine and tryptophan fluorescence are blue-shifted, if the amino acids move into a more hydrophobic (non-polar) environment and red-shifted if the amino acids move into a more hydrophilic (polar) environment (*Fig. 10.7*).

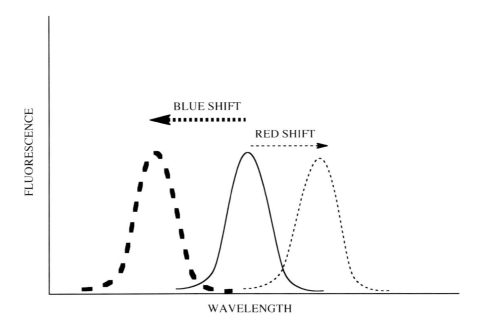

Figure 10.7: Three overlaid fluorescence spectra illustrating a typical spectrum and the impact of a blue shift or a red shift.

Note the change in peak λ when more levofloxacin is added for both Titration #1 and Titration #2. What can you conclude about the changing environment around the fluorophores?

PROCEDURES

Reagents and equipment needs are calculated per six student teams. There is ~20% excess included.

Equipment/glassware needed:
1. Three sets of micropipettes 20-100 μl, 2-20 μl and 1-10 μl
2. standard cuvette for a spectrofluorometer

Reagents needed:
1. 10 ml of 0.014 M levofloxacin in 0.05 M tris, pH = 7.0
2. 100 ml of 1 x 10^{-5} M serum albumin in 0.05 M tris, pH = 7.0

Experimental procedure:

Titration #1:
1. Set spectrofluorometer to $\Delta\lambda$ = 15 nm.
2. Pipet 2.0 ml of bovine serum albumin (1 x 10^{-5} M; 0.05 M tris, pH = 7.0) into the fluorescence cuvette. Measure synchronous spectra from 250 nm to 450 nm.
3. Add 1 μl of 0.014 M levofloxacin (0.05 M tris, pH = 7.0). Measure synchronous spectra from 250 nm to 450 nm.
4. Repeat the levofloxacin additions followed by synchronous fluorescent measurements two more times.
5. Add 2 μl of 0.014 M levofloxacin. Measure synchronous spectra from 250 nm to 450 nm.
6. Repeat the levofloxacin additions followed by synchronous fluorescent measurements four more times.
7. You should have eight different spectra at this point. Plot data as follows:
a. Create an overlay of all spectra.
b. Record the fluorescence at an excitation λ = 287 nm for each spectra.
c. Print overlay spectra.
8. Rinse the cuvette carefully with distilled water. The albumin solution may be discarded in the sink.

Titration #2:
1. Repeat the same procedure, but set the spectrofluorometer to $\Delta\lambda$ = 60 nm.
2. Rinse the cuvette carefully with distilled water. The albumin solution may be discarded in the sink.

Data Analysis:

1. Calculate fluorescence quenching:

$$\text{Quenching} = \frac{\Delta Fl}{\Delta Fl_{initial}} = \frac{Fl_{initial} - Fl}{Fl_{initial}} \qquad (7)$$

Note that fluorescence quenching varies from 0 to 1.

2. Calculate the concentration of levofloxacin in the cuvette; [levofloxacin]$_{cuvette}$ for each addition. Remember that the levofloxacin is *diluted* when added to the cuvette. You need to use the dilution equation, $M_1V_1 = M_2V_2$, where M_1 = stock concentration of levofloxacin, V_1 = total volume of levofloxacin added to the cuvette, M_2 = the cuvette concentration of levofloxacin, and V_2 = the total volume in the cuvette.

3. Plot the fluorescence quenching versus [levofloxacin]$_{cuvette}$ for Titration #1 and Titration #2.

4. Determine K_D values for Titration #1 and Titration #2 as follows.

a. Plot $\dfrac{Fl_{initial}}{Fl}$ versus $[Levofloxacin]_{curvette} \dfrac{Fl_{initial}}{\Delta Fl}$ for Titration #1 and Titration #2. Note that we are using the linear form of the K_D equation (see Equation 5).

b. $K_D = \dfrac{1}{slope}$ of this graph.

Notes to the Instructor
The experiment in Chapter 10 uses the versatility of a research-grade spectrofluorometer. The spectrofluorometer gives a complete view of the observed spectral changes. Students measure complete spectra and then analyze fluorescence at specific wavelengths. Synchronous spectroscopy is possible with such a spectrofluorometer. A more basic fluorometer with filters rather than monochromators can also be used. In this case, the student should be provided with the pertinent spectra.

Albumin - Levofloxacin Binding
Lab Report Outline and Point Distribution

Introduction
1. Several sentences defining the goal/purpose of this experiment. (2 pts.)

Data
1. A table reporting the fluorescence as a function of the total volume of levofloxacin solution added for Titration #1. (2 pts.)
2. A table reporting the same data for Titration #2. (2 pts.)

Results
1. An example of the calculation you used to find the levofloxacin concentrations in the cuvette. (2 pts.)
2. Four plots:
   a. Fluorescent Quenching vs. $[\text{levofloxacin}]_{\text{cuvette}}$ for Titration #1 and #2. (4 pts. each)
   b. $\frac{Fl_{initial}}{Fl}$ vs. $[\text{levofloxacin}]_{\text{cuvette}}$ $\frac{Fl_{initial}}{\Delta Fl}$ for Titration #1 and #2. Determine the K_D from each plot. (5 pts. each)

Analysis
1. Access the fluorescence excitation and emission spectra for phenylalanine, tyrosine and tryptophan using the Internet. Answer the following four questions. (8 pts.)
   a. Estimate peak wavelengths for absorption (at $\lambda > 230$nm) and emission (at $\lambda > 250$nm) for each amino acid.
   b. How are the peak absorption wavelengths related to the peak emission wavelengths? For example, is the emitted light at lower or higher energy than the absorbed light? Briefly explain.
   c. Which amino acid shows the biggest energy change from absorption to emission? Explain your reasoning.
   d. Rank these three amino acids from most fluorescent to least fluorescent. Briefly explain your reasoning.
2. Identify the titration (either $\Delta\lambda = 15$ nm or $\Delta\lambda = 60$nm) that monitors tyrosine fluorescence and the titration that monitors tryptophan fluorescence. Which amino acid accounts for most of albumin fluorescence? Briefly explain. Which amino acid is most quenched by levofloxacin binding? Briefly explain. (6 pts.)
3. Is there evidence for an albumin conformation change caused by levofloxacin binding? If there is evidence for a conformation change, are the tyrosines changing their environment (becoming more exposed to solvent or more buried into the protein interior)? Answer the same question for the tryptophans. Briefly explain being as specific as possible. (10 pts.)

11 Understanding the Importance of Buffers in Biological Systems

11.1 Learning Objectives

You should be aware that buffers play a critical role in almost all biochemical systems. Biochemical experiments routinely require a buffer. In this laboratory you will cover the basics of buffer preparation and test the buffering capacity of the resulting solution. This buffer will be used in the enzyme kinetics (acetylcholinesterase) lab later in the term.

11.2 Introduction

Most biochemical reactions either produce or consume hydrogen ions. Take, for example, the reaction catalyzed by acetylcholinesterase. This enzyme rapidly destroys the neurotransmitter acetylcholine (Ach) after it has delivered its signal across synapses such as the neuromuscular junction. The acetylcholinesterase reaction hydrolyzes acetylcholine, to choline, acetate and a hydrogen ion. Note the hydrogen ion liberated by this reaction.

Typically, during nerve signaling the synaptic [Ach] will increase from zero to the millimolar concentration range. Destroying this acetylcholine will release millimolar quantities of hydrogen ions. If uncontrolled, this simple reaction would decrease the pH from about 7 to below 6. This pH change is far beyond anything that human cells can tolerate. Similar wide swings in pH can arise from almost every metabolic process. So, pH control is necessary for life; this control is provided by buffers.

11.3 Buffer Preparation

Buffers are an aqueous solutions of weak acids or bases that minimize a pH change. Because these acids/bases are "weak," they establish an equilibrium in solution

$$HA \rightleftharpoons H^{\oplus} + A^{\ominus}$$

that can be described with the equation,

$$K_a = \frac{[A^-][H^+]}{[HA]}$$

(1)

Rearranging Equation 1:

$$[H^+] = K_a \frac{[HA]}{[A^-]}$$

(2)

Equation 2 shows that the hydrogen ion concentration depends on the conjugate acid concentration [HA] and the conjugate base concentration [A⁻] as well as the equilibrium constant K_a. That is, the [H⁺] is under control.

Because pH is commonly measured instead of [H⁺], Equation 3 is most often presented in a modified form (the *Henderson-Hasselbalch Equation*):

$$pH = pK_a + log \frac{[A^-]}{HA} \text{ where } pH = -log[H^+] \text{ and } pK_a = -logK_a$$

(3)

This equation emphasizes that pH changes as the ratio of concentrations of A⁻ to HA changes. So, adjusting the concentration of [HA] and [A⁻] will set the pH.

For example:

– In pure water, the pH is 7. Adding HCl (a strong acid) to 0.001 M will increase the [H⁺] to about 0.001 M. The pH will change to about 3, a difference of four pH units.
– In a buffer solution, the same addition of HCl will cause a shift in the buffer equilibrium, rather than a drastic pH change. For example, let [A⁻] = [HA] = 0.1 M before adding HCl.
   - Adding 0.001 M HCl will convert 0.001 M A⁻ to 0.001 M HA.
   - The buffer concentrations will change, [A⁻] to 0.099 M and [HA] to 0.101 M.
   - The $log\frac{[A^-]}{HA}$ will change from $log\frac{[0.1\ M]}{[0.1\ M]} = 0$ to $log\frac{[0.099\ M]}{[0.101\ M]}$. According to the Henderson-Hasselbalch Equation, the pH will decrease by 0.009 units.
– Instead of reacting with water and changing the pH by four units, the HCl reacts with the buffer and only changes the pH by 0.009 units.

Four important generalizations about buffers

1. A buffer is composed of an equilibrium mixture of a weak acid (HA) and its conjugate weak base (A-).
2. The higher the buffer concentration, the greater the pH control.
3. No matter what the buffer concentration, maximum pH control is reached when [HA] = [A-]. At this point, the Henderson-Hasselbalch Equation gives pH = pKa + 0. So, the maximum pH control occurs when the pH is numerically equal to the pKa.
4. It is good practice to choose a weak acid whose pKa is close to the pH you are targeting. Typically, weak acids are effective buffers at pHs within one unit of their pKa.

Procedures

Reagents and equipment needs are calculated per six student teams. There is ~20% excess included.

Equipment/glassware needed
1. six 100 ml beakers
2. six stir bars
3. six stir plates
4. six 100 ml graduated cylinders
5. six pH meters of same kind
6. small box of plastic Pasteur pipets

Solutions needed
1. 250 ml 0.5 M HCl
2. solid tris HCl
3. solid tris base
4. pH 7.0 standard buffer
5. pH 4.0 standard buffer

I. Using the conjugate acid, trisHCl, and the conjugate base, tris, prepare one hundred milliliters of a 0.2 M buffer at pH = 8.0. (Tris has a pKa = 8.1.)

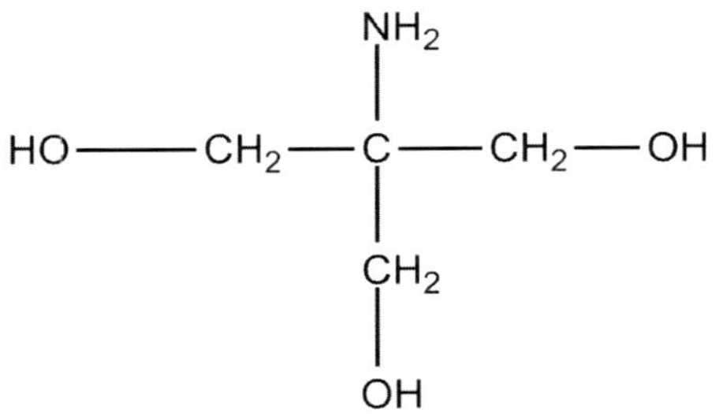

$$HO——CH_2—\overset{\overset{\displaystyle NH_2}{|}}{\underset{\underset{\displaystyle OH}{|}}{\underset{CH_2}{|}}{C}}—CH_2—OH$$

1. Using the Henderson-Hasselbalch equation, determine the concentrations of both $[A^-]$ = tris = x and [HA] = trisHCL = (0.2 - x) $[A^-]$.
2. Calculate masses needed.
3. Add the masses to a 100.00 ml volumetric flask and fill to the mark with "Millipore" water.
4. Transfer the stock buffer to a bottle, appropriately labeled with the contents (0.2 M tris buffer), the pH, your name(s) and date.

II. Using a pH meter, determine buffer pH and buffering capacity.

A pH meter is a common piece of laboratory equipment that requires some care in use. The glass bulb on the end of the electrode is fragile and easily broken. Always rinse the electrode with distilled water when moving from one solution to another. The pH meter should be set in standby mode when the electrode is out of solution.

A pH meter offers a *relative* measure of pH and, therefore, must be standardized. Typically, two standard buffers are used. A first buffer (commonly, pH=7.0) is used to make major adjustments; then, a second buffer (pH=4.0) is used to make fine adjustments. The pH meter will have two different dials - one for major adjustments and one for fine adjustments.

1. Measure the pH of your buffer solution.
2. Prepare 32 ml of a 1:4 dilution of your buffer. (Graduated cylinders may be used for the volume measurements.) Measure the pH of this dilution.
3. Add 0.5 M HCl dropwise to drive the pH down below 6.
4. Titrate the buffer solution back to pH 11 with 0.1 M NaOH. Record about twenty points to graph your titration curve later. Your burette readings should be taken to the nearest 0.05 ml.
5. Prepare 32 ml of a 1:40 dilution of your buffer. (In addition to a graduated cylinder, use a pipette for this dilution.) Measure the pH of this dilution.
6. Add 0.5 M HCl dropwise to drive the pH down below 6.
7. Titrate the buffer solution back to pH 11 with 0.1 M NaOH. Record about twenty points to graph your titration curve later. Your burette readings should be taken to the nearest 0.05 ml.
8. Store your buffer at 4 °C for use with the acetylcholinesterase kinetics laboratory later in the semester.

III. Graph the titrations.

1. Using a computer graphing program, plot pH versus mmol of OH^- for both titrations. Note the impact of tris on the pH change.
2. Qualitatively, buffering capacity can be defined as the amount of strong acid/base that can be added to a buffer solution before causing a significant pH change. Buffering capacity can be quantified by taking the inverse of the instantaneous slope of pH vs. OH^- amount. A quick, empirical approach to measuring the buffering capacity is to note the amount of NaOH needed to change the pH by one unit in the middle of the buffering region. Using this last method, determine the buffering capacity for your tris solution from each titration.

Notes to the instructor

The tris/trisHCl buffer was chosen to accommodate the acetylcholinesterase enzyme kinetics lab later in the term. Any buffer system may be substituted. From a pedagogical perspective, a follow-up lab that uses the buffer to study a biochemical reaction is appropriate.

Prelab for the Buffer Lab

1. (2 pts.) Describe the reaction catalyzed by acetylcholine esterase!

2. (2 pts.) Define/describe buffer.

3. (6 pts.) During the lab you will prepare 100 ml of Tris buffer at pH=8.0 using the conjugate acid, trisHCl, and the conjugate base, Tris, (Tris has a pKa =8.1.).

a. Using the Henderson-Hasselbalch equation, determine the concentrations of both [A-] = tris = x and [HA] = trisHCl = (0.2 - x).

b. Calculate masses as needed.

Buffer Lab Report
Outline and Point Distributions

Introduction:
1. Write several sentences defining the goal/purpose of this experiment (3 pts.).

Data:
1. Show the calculations you used to prepare the 0.2 M tris buffer (3 pts.).
2. Show the calculations you used to prepare the 1:4 and 1:40 diluted tris buffer (3 pts.)
3. Report the pH for all three buffers (3 pts.).

Results:
1. Graph both titrations using a computer graphing program (3 pts./graph).
2. Show how you determined the pK_a and the buffering capacity on your graph for each titration (4 pts.).
3. Report the buffering capacity and pK_a as determined for each titration (4 pts.).

Analysis:
1. Suggest some possibilities as to why your buffer pH might not be exactly 8.0 (3 pts.).
2. Does the center point of the tris buffering region ("best buffering") match the pK_a? Briefly explain (3 pts.).
3. Does the buffering capacity change in a predictable way? Briefly explain (3 pts.).

Questions/Problems:
1. Include the solutions to the Buffer Problems handout (15 pts.).

Buffer Problem Set

The majority of buffer problems can be solved using one equation, the Henderson-Hasselbalch equation:

$$pH = pK_a + \log \frac{[A^-]}{HA} = pK_a + \log \frac{[\text{conjugate base}]}{[\text{conjugate acid}]}$$

1. Calculate the pH of a one liter solution containing 0.15 mol benzoic acid and 0.25 mol sodium benzoate. The pK_a for benzoic acid is 4.2.

The next set of problems illustrates the two common methods to prepare a buffer.

2. Add a known amount of conjugate acid to a known amount of conjugate base: What is the pH of 0.5 l of a buffer prepared by mixing 8.6 g of lactic acid (90 g/mol) with 7.8 g of sodium lactate (112 g/mol)? The pKa for lactic acid is 3.86.

3. Calculate the mass of benzoic acid and sodium benzoate (in grams) needed to prepare 250 ml of a 0.1 M buffer at pH = 4.1. (The buffer concentration is defined as the sum of the conjugate acid concentration plus the conjugate base concentration.) The pKa of benzoic acid is 4.2.

4. Start with either the conjugate acid or base and add a strong base or strong acid (conjugate acid plus strong base forms the conjugate base and water; conjugate base plus strong acid forms the conjugate acid): What is the pH of 0.5 l of a 0.1 M acetic acid solution to which 0.73 g of NaOH are added? The pKa of acetic acid is 4.76.

5. You are asked to prepare 1.2 l of a 0.05 M tris buffer at pH= 7.8. You start with the conjugate base form of tris (121 g/mol). How many grams of tris must you weigh out? How many ml of 6 M HCl (a strong acid) must you add to reach pH=7.8? The pKa for tris is 8.1.

12 Molecular Visualization of an Enzyme, Acetylcholinesterase

12.1 Learning Objectives

The goal of this laboratory is to analyze some of the major structural elements of an important enzyme, acetylcholinesterase (AChE). To do this, you will use a common structural visualization program and correlate AChE structural elements with the enzyme mechanism. You will be using Chimera, a state-of-the-art molecular visualization program provided by the National Science Foundation through the University of California, San Francisco. This free program is available at http://www.cgl.ucsf.edu/chimera/.

To provide a quick overview of the program, we will look at a multi-drug resistance efflux pump - the protein that is controlled by the riboswitch you studied. You will then use this program to analyze the enzyme, acetylcholinesterase.

12.2 Introduction and Background

Acetylcholinesterase (AChE) destroys the nerve transmitter, acetylcholine by hydrolysis.

AChE is one of the most efficient enzymes in nature - in some ways, a "perfect" enzyme. Amino acid side chains at the active site are precisely arrayed to force bonding changes in the acetylcholine (*Fig. 12.1*).

Overall, the electrons are pushed toward the ester carbonyl forming a covalent intermediate between the reactant and the enzyme (*Fig. 12.2*).

Acetylcholinesterase's finely tuned mechanism provides a good illustration of some common catalytic processes:

- Covalent Intermediate Catalysis - The activation energy is lowered because an especially reactive active site group attacks and forms a new covalent bond to the reactant;
- Acid-Base Catalysis - The activation energy is lowered because active site groups force electron (and proton) movement.
- This week you will examine this enzyme at a molecular level. Next week you will study the enzyme kinetics of acetylcholinesterase.

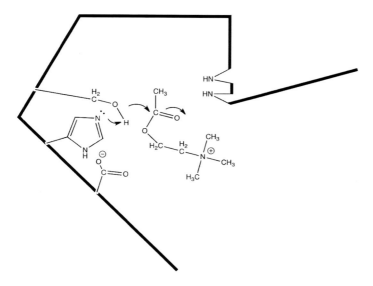

Figure 12.1: A schematic of the acetylcholinesterase active site showing the electron movement during the initial nucleophilic attack of acetylcholine: The alcohol acts as a nucleophile attacking the ester carbonyl carbon. This alcohol is a strong nucleophile because the adjacent imidazole ring changes the alcoholic hydroxyl group to an alkoxide. In other words, the imidazole acts as an especially strong base. It can act in this way, because of the adjacent carboxylate negative charge.

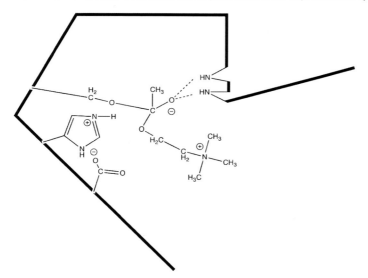

Figure 12.2: A schematic of the acetylcholinesterase active site showing the covalent intermediate formed following the initial nucleophilic attack: Note that the negative charge that forms on the carbonyl oxygen is stabilized by hydrogen bonding with "-NH" groups. Once this covalent intermediate forms, choline rapidly leaves the active site and water enters to complete the reaction.

12.3 Introduction to Molecular Visualization Using the Program Chimera

The visualization program Chimera will be introduced to you by using the structure of the multi-drug resistance (MDR) efflux pump protein. This is the protein that is up-regulated by a riboswitch in response to antibiotics like tetracycline. The protein then pumps the antibiotics out of the bacteria allowing the cells continued growth. Because the MDR efflux pump eliminates many different drugs, the bacteria become *multi-drug* resistant.

The MDR efflux pump is an integral membrane protein. As such, the protein structure has some common membrane protein structural characteristics. First, transmembrane helices comprise the bulk of the protein. These helices stretch the width of the lipid bilayer and expose hydrophobic amino acids to the lipid bilayer interior. Hydrophilic amino acids are located where the protein meets the lipid bilayer surface, both on the outside of the cell as well as on the inside.

The instructions for using Chimera follow this format:
    Italics   indicates that you should go to a *Menu* or a *Command* line or the *Cursor*
    Bold... tells you which menu to access
    ...Bold indicates a selection from the accessed menu
    Bold   indicates a command

1. First, download the protein data. Structural data for macromolecules is available from a central data bank, the RCSB Protein Data Bank. Every structure is given its own unique code. For example, the multi-drug resistance efflux pump data set used here is "2GFP."
    Menu: **File... Fetch by ID**
        type in box 2GFP
        click **Fetch** button

2. The command line can be used for very specific changes. To bring up the command line (or hide it) go to the "Tools" menu, "General Controls" sub-menu.
    Command: **del :**.b
        Command: **focus**

3. The cursor (mouse) provides for quick changes in the protein view.
    Cursor: left mouse button + moving cursor = rotates protein
    Cursor: ctrl + left mouse button + moving cursor = selects protein
    Cursor: right mouse button + moving cursor = changes size of protein
    Cursor: ctrl + right mouse button + moving cursor = translates protein
    Place cursor over the protein to identify specific amino acids

4. The "Presets" menu gives choices for common ways to represent the protein.
   Menu: **Presets... Interactive 1**
     Menu: **Presets... Interactive 2**
     Menu: **Presets... Interactive 3**
     Menu: **Presets... Interactive 4**

5. The "Select" menu specifies what part of the protein will be changed by the "Actions" menu. Try the following three examples:

   a. Example #1: A chain of amino acids linked by peptide bonds (a polypeptide) is selected and any action now will apply to that chain (e.g., changing color).
      Menu: **Select... Chain...** A
      Menu: **Actions... Color...** ?

   b. Example #2: All amino acids that carry a negatively charged side chain at neutral pH (i.e., carboxylates) are selected and all the atoms/bonds in the side chains are (a) shown and (b) colored by element.
      Menu: **Select... Clear Selection**
      Menu: **Select... Residue... amino acid categories... negative**
      Menu: **Actions... Atoms/bonds... show**
      Menu: **Actions... Color... by element**

   c. Example #3: All amino acids that carry a positively charged side chain at neutral pH are selected and all the atoms/bonds in the side chains are shown and colored by element. The atoms are then shown as actual size (sphere). Finally, the protein is shown as a solid object with a surface, as it would actually appear.
      Menu: **Select... Selection Mode (replace)... append**
      Menu: **Select... Residue... amino acid categories... positive**
      Menu: **Actions... Atoms/bonds... show**
      Menu: **Actions... Color... by element**
      Menu: **Actions... Atoms/Bonds... sphere**
      Menu: **Actions... Atoms/Bonds... hide**

      Menu: **Select... Clear Selection**
      Menu: **Select... Chain...** A
      Menu: **Actions... Surface... show**

6. The tools menu can provide further information about protein properties.
   Menu: **Tools... Surface/Binding Analysis... Coulombic Surface Coloring**
        click **OK** *button*
   Menu: **Actions...Surface...Hide**

Menu: **Tools... Depiction... Color Secondary Structure**
    click **OK** *button*
Menu: **File... Close Session**

12.4 Analysis of Acethylcholinesterase Using the Computer Visualization Program Chimera

The following series of tasks help you learn how to use computer visualization software to better understand how the enzyme acetylcholinesterase works. Follow the instructions below to answer the questions.

1. The overall structure of acetylcholinesterase. Proteins are stabilized by secondary structures, commonly either β-pleated sheets or α-helices.

Menu: **File... Fetch by ID**
    type in box 1AMN
    click **Fetch** button
Menu: **Tools... Depiction... Color Secondary Structure**
    click **OK** *button*
Command: **del** :HOH :SO4

a. (2 pts.) How many strands are included in each of the two β-pleated sheets?

b. (3 pts.) Identify the two longest α-helices by listing the abbreviations for the amino acids at the beginning and the end of each helix. (Placing your cursor over a spot on the protein will cause the abbreviation to be shown.) How many amino acid residues are in each helix? [Hint: the amino acids are numbered consecutively.]

2. Substrate Analog (NAF) at the Active Site. Active sites are often marked or labeled with a substrate analog. This is a substrate-like molecule that reacts incompletely at the enzyme active site. It remains bound to the enzyme and marks some of the catalytic amino acid side chains. In this case the substrate analog forms a structure like the tetrahedral intermediate shown in the Introduction and Background.

> *Menu:* **Select... Residue... NAF**
> *Menu:* **Actions... Atoms/Bonds... Show**
> *Menu:* **Select... Invert (all models)**
> *Menu:* **Actions... Ribbon... Hide**
> *Menu:* **Actions... Atoms/Bonds... Hide**
> *Command:* **display: 200**
> *Command:* **focus**

a. (4 pts.) Draw the structural formula for NAF connected to an atom of an amino acid side chain (treat the rest of the side chain as an "R" group). This structure contains two ionic charges (one positive and one negative). Can you locate these charges? Mark each with either a "+" or a "-."

b. (4 pts.) NAF forms a covalent intermediate with the enzyme just as the natural substrate, acetylcholine, does. This intermediate is called a "tetrahedral intermediate." Why do you think it is given that name? Be as specific as possible.

3. Noncovalent Interactions between the Active Site and the Substrate Analog. Van der Waals contacts between the protein and the substrate analog are very common.

> *Menu:* **Select... Clear Selection**
> *Menu:* **Select... Residue... NAF**
> *Menu:* **Tools... Structure Analysis... Clashes/Contacts**
>   *click* **Designate** button
>   *click* **Contact** button
>   *check* **Select**
>   *uncheck* **Draw pseudobonds of color**
>   *check* **Color**
>   *click* **OK** button

a. (3 pts.) How many atoms of NAF are directly in contact with the protein? Out of a total of how many atoms in NAF? [Atoms marked in red are in direct contact with the protein.] Mark each atom that is in contact with an asterisk (*) in your NAF structure in question #3.

4. Visualize the amino acids that immediately surround the substrate analog *click on Graphics Window then hit the arrow up key* [This will select the complete NAF.]

> *Menu:* **Actions... Atoms/Bonds... Show**
> *Menu:* **Actions... Atoms/Bonds... Sphere**
> *Command:* **focus**
> *Menu:* **Select... Clear Selection**
> *Menu:* **Select... Residue...** NAF
> *Menu:* **Actions... Color...** choose a color that is easy to see

a. (4 pts.) Does the NAF have any "wiggle room" when bound at the active site? How do you think that this might aid enzyme catalysis?

5. Select active site amino acids that help catalyze the reaction.

> *Menu:* **Actions... Atoms/Bonds... Stick**
> *Menu:* **Actions... Atoms/Bonds... Color... by element**
> *Menu:* **Select...Invert (all models)**
> *Menu:* **Actions... Atoms/Bonds... Stick**
> *Menu:* **Actions... Atoms/Bonds... Color... by element**
> *Command:* **show** :118 :119 :200 :327 :440 :NAF
> *Command:* **focus**

a. (3 pts.) Identify the α-carbon atom for each amino acid. The α-carbon lies between a carboxyl "C=O" and an "N-H" along the protein backbone. The side chain for each amino acid is also connected to the α-carbon. (The side chain for a glycine amino acid is simply a "H" and so it is not shown in the structure.) One side chain is covalently bonded to NAF. Move your cursor to the α-carbon to learn the abbreviated name for this amino acid. Use the list of amino acid structures below to identify this amino acid and draw its side chain structure.

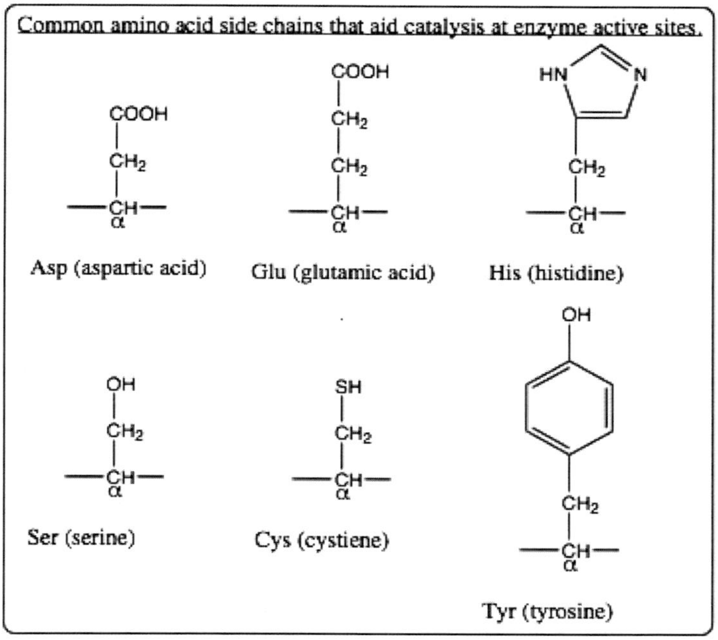

Common amino acid side chains that aid catalysis at enzyme active sites.

b. (4 pts.) Draw structures for all other amino acid side chains shown at this active site. Identify each amino acid as well. At the catalytic step shown by this structure determined by x-ray crystallography, two of the amino acid side chains carry an ionic charge (one positive and one negative). Identify the charged side chains and give them their appropriate charges.

Menu: **Tools...Structural Analysis... FindHbond**
        click **OK** button

c. (3 pts.) The blue lines show H-bonds and the H-bonded atoms are close enough together to catalyze reactions.

Two hydrogen bonds that connect to NAF are <u>not</u> from amino acid side chains. Describe or identify the groups from which these H-bonds originate.

Measure the distance between atoms along the blue lines (H-bonds). [Place the cursor over a H-bond. A box will appear detailing the atom within each residue that is involved in H-bonding as well as the distance of the H-bond.] Identify the atoms and give distances.

Ctrl + Left click and drag over all atoms to select
Menu: **Actions... Atoms/Bonds... sphere**

d. (4 pts.) Note that where H-bonding occurs, the electron clouds are overlapping. What does this mean in terms of bonding?

6. Location of the Active Site with respect to the Overall Protein Structure - the active site can often be recognized as a "pit" or "crevasse" or an "indentation" in the enzyme surface.

Menu: **Select... Chain... A**
Menu: **Actions... Ribbons... Show**
Menu: **Select... Clear Selection**
Command: **focus**
Menu: **Actions... Surface... Show**
Menu: **Select... Residue... NAF**
Menu: **Actions... Colors... green**

a. (4 pts.) Locate NAF. This is a marker for the location of the active site. Describe the active site location relative to the bulk protein. Might this location cause problems for the enzyme? Briefly explain.

7. Protein Surface Charges and Catalysis

  Command: **del:NAF** [This will leave a "hole" in the active site.]

  Menu: **Tools... Surface/Binding Analysis... Coulombic Surface Coloring**

    choose 11 for number of colors/values

    click **OK** button

    [Close any error messages. This calculation may take a minute.]

a. (3 pts.) Describe the location of the predominately blue region (positively charged region) with respect to the active site. How might this relative location help catalysis?

b. (3 pts.) Describe the location of the predominately red region (negatively charged region) with respect to the active site. How might this relative location help catalysis?

8. Protein Motion During Catalysis - proteins are commonly pictured as static, but they flex and bend as part of their function.

  Menu: **File... Close Session**

  Menu: **File... Fetch by ID**

    type in box 2v96

    click **Fetch** button

  Menu: **Presets... Interactive 1 (ribbons)**

  Command: **del :.b**

Menu: **File... Fetch by ID**
  type in box 2va9
  click **Fetch** button
Command: **del :.b**

Menu: **Tools... Structure Comparison... Match Maker**
  highlight reference structure 2V96
  check **After superposition, compute structure-based multiple sequence alignment**
  click **Apply** button
  check **Iterate superposition/alignment...** in window that opens
  check **Iterate alignment until convergence**
  click **OK** button

Menu: **Tools... Structure Comparisons... Morph Conformations**
  click **Add** button
  double click on 2V96.pdb (#0) *in new window*
  double click on 2VA9.pdb (#1) *in new window*
  double click on 2V96.pdb (#0) *in new window*
  select **Action on Create: hide Conformations**
  click **Create** button
Command: **focus**
  MD Movie: Molecular Movement... [This is a new window that opens.]
  click **Play** button

a. (3 pts.) Can you find motion in this protein as it catalyzes the reaction? What type of structures move - a loop, an α-helix, a β-pleated sheet?

9. Protein Motion at the Active Site

Command: **display** #2 :200
(Note that this marks the active site using one of the active site amino acids.)
Command: **display** #2 :70 :74 :84 :121 :279 :290 :330 :331 :334
(Note that these amino acids line the "tunnel" into the active site.)

a. (3 pts.) These motions have been described as "breathing" of the "tunnel." Why would you expect motions like this in the tunnel that leads to the active site? Briefly explain.

Notes to Instructor

This laboratory allows the students to "try out" the Chimera program - to experiment with different settings and commands. We find that Chimera is relatively user-friendly and forgiving. Students are encouraged to download this free software for their own use. Chimera may be used with at-home assignments based on the lecture portion of a biochemistry course as well. This is a good opportunity to connect with topics covered in lecture as well as the laboratory.

Molecular Visualization of Acethylcholinesterase Prelab

Using any common sources (internet, biochemistry textbook, etc.), answer the following questions concerning the general characteristics of the enzyme, acetylcholinesterase.

1. (3 pts.) What reaction is catalyzed by acetylcholinesterase? Show the structures of reactants and products.

2. (3 pts.) In general terms how is acetylcholinesterase important to neuronal transmission across a synapse?

3. (3 pts.) Where is acetylcholinesterase located within the synaptic space? For example, is acetylcholinesterase found on the presynaptic membrane or on the postsynaptic membrane or as a soluble enzyme in the synaptic cleft or ... ? In general, is this enzyme free to diffuse away from the synapse? Briefly explain.

Acetylcholinesterase Characteristics Worksheet

Using any common sources (internet, biochemistry textbook, etc.), answer the following questions concerning the general characteristics of the enzyme, acetylcholinesterase.

1. (2 pts.) Is the natural form of acetylcholinesterase a single polypeptide chain (a monomer)? Or is acetylcholinesterase found as a polymer? Explain.

2. (2 pts.) Briefly describe the steps acetylcholine takes, starting from being in the presynaptic vesicle and ending when acetylcholine is reacted by acetylcholinesterase.

3. (3 pts.) Many nerve gasses (e.g., sarin (GB), soman (GD)) were designed to impact acetylcholinesterase. Specifically, what do these nerve gasses do to this enzyme? How does the change in acetylcholinesterase (brought about by nerve gasses) affect nerve transmission?

4. (3 pts.) Many common insecticides (e.g., malathion, Sevin) also affect acetylcholinesterase. Also some drugs used to treat diseases such as Alzheimer's and myasthenia gravis target acetylcholinesterase. Examples of these drugs include physostigmine (eserine), neostigmine, and pyridostigmine. How is the activity of acetylcholinesterase impacted by these insecticides or drugs? How does this change in acetylcholinesterase affect nerve transmission?

13 Determining the Efficiency of the Enzyme Acetylcholine Esterase Using Steady-State Kinetic Experiment

13.1 Learning Objective

This laboratory introduces you to steady-state kinetic analysis, a fundamental tool for studying enzyme mechanisms. The enzyme studied, acetylcholinesterase (AChE), has a well-understood mechanism and carefully examined structure. Additionally, AChE is physiologically very important and is an example of "catalytic perfection." You will determine V_{max}, k_{cat}, and K_M and then analyze the catalytic capabilities of AChE.

13.2 Measuring the Catalytic Efficiency of Acetylcholinesterase

Acetylcholinesterase catalyzes the hydrolysis of the neurotransmitter acetylcholine to choline and acetate:

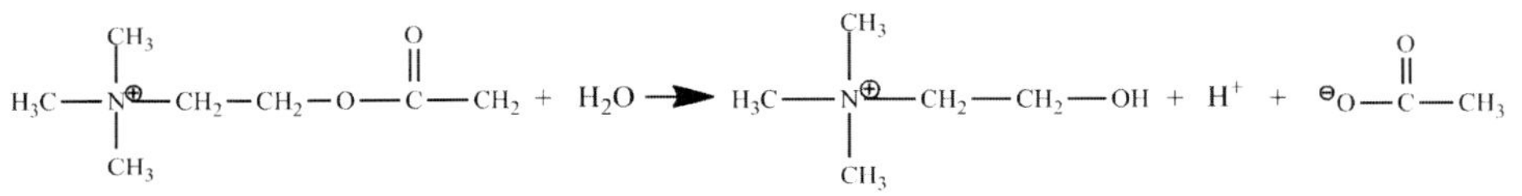

The organic reactant in an enzymatic reaction is also termed the *substrate*. By studying the kinetics of this reaction, you will probe the details of acetylcholinesterase catalysis.

As is the case with many enzyme studies, you will use a different substrate that gives a color change upon reaction, acetylthiocholine (ASCh). A sulfur atom replaces the oxygen atom in this compound's ester.

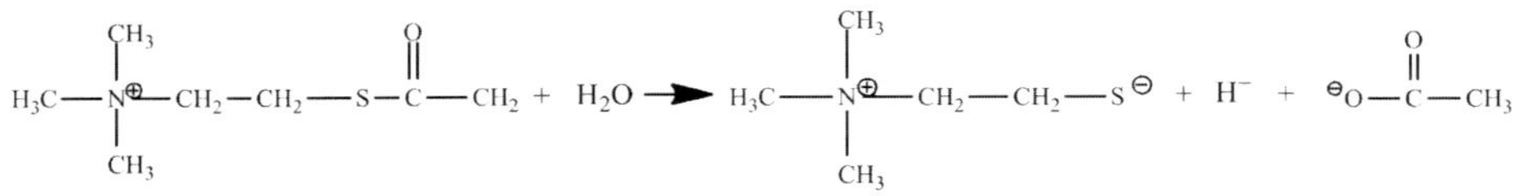

ASCh reacts like acetylcholine, but one of its products is different. This different product, thiocholine, reacts with the color-forming reagent, 5,5´-dithiobis(2-nitrobenzoate) (DTNB) producing a strong yellow color.

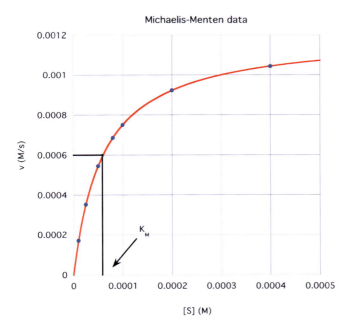

13.3 Running a Steady-State Enzyme Kinetics Experiment

The most common steady-state enzyme experiment holds the enzyme concentration constant and measures reaction rate (velocity, v) at varying reactant (substrate, S) concentrations. Velocity is the *dependent* variable and substrate is the *independent* variable. This experiment generates the Michaelis-Menten plot. An example follows (*Fig. 13.1*):

Michaelis-Menten data

Figure 13.1 : A Michaelis-Menten plot graphs steady-state reaction velocity as a function of substrate concentration: Each data point on the plot represents one assay. This plot reports data from seven assays.

13.4 Designing a Steady-State Experiment

Commonly, an assay solution contains buffer, substrate, enzyme and often a color-forming reagent. Your assays will start with a milliliter volume of buffer into which microliter volumes of stock substrate, stock color-forming reagent, and stock enzyme are diluted. To prepare for a steady-state enzyme kinetics experiment you have to fill out a chart similar to the one below.

Table of Volumes for Each Assay

Assay #	$[ASCh]_{assay}$	Buffer volume	ASCh volume	DTNB volume	Enzyme volume	Total assay volume
1						
2						
etc.						

Instructions on how to calculate each quantity are as follows:

1. [ASCh]assay:

Refer to the Michaelis-Menten plot on *Fig. 13.1*. The Michaelis constant (K_M) is related to the attraction between the enzyme and the substrate. That is, a smaller K_M means the enzyme reacts at smaller substrate concentrations - the substrate is more strongly attracted to the enzyme.

As shown, the K_M is also a rough midpoint on the hyperbolic curve. Commonly, about one half of the assays should have [S] < K_M and about one half of the assays, [S] > K_M. Typically, the largest [S] is chosen to be about 5 x K_M. Because of the hyperbolic curve shape, the smaller [S] are spaced closer together while the larger [S] values are spaced further apart. The K_M provides a guide for choosing substrate concentrations.

Acetylcholinesterase (AChE) has a K_M for the substrate (acetylthiocholine iodide, ASCh, M.W. = 289.18 g/mol) of about 8 x 10^{-5} M. Choose seven substrate concentrations up to 5 x K_M being careful to space the concentrations appropriately. Fill in the column marked "$[ASCh]_{assay}$" in your Table of Volumes for Each Assay.

Understand that the concentrations you have just chosen are under assay conditions - that is, <u>in the cuvette.</u> You get these cuvette concentrations by adding small volumes of a stock [ASCh] to a larger volume of buffer (a dilution). You now need to calculate (a) the stock [ASCh] concentration and (b) the microliter volumes of this stock [ASCh] to be used for each of these assays.

2. Buffer volume:

You will use a 1:4 dilution of the tris buffer you prepared several weeks ago in a 3.0 ml assay volume. For the experiments you will use standard quartz cuvettes.

3. Stock [ASCh] concentration:

When you calculate the stock [ASCh] concentration ([ASCh]$_{stock}$) you need to consider two things: (1) The highest [ASCh] you want to use in the assay. As we mentioned earlier, this is typically 5 x K$_D$ value of the enzyme; (2) [ASCh]$_{stock}$ gets diluted when it is added to the buffer in the cuvette. Consequently, you need to use the dilution equation:

$$M_1V_1 = M_2V_2 \tag{1}$$

where:

> M$_1$ is [ASCh]$_{stock}$
> V$_1$ is 100 μl for a 3.0 ml assay
> M$_2$ is [ASCh]$_{assay}$
> V$_2$ is the total volume in the cuvette.

$$V_2 = V_{buffer} + V_{enzyme} + V_{DNTB} + V_{ASCh} \tag{2}$$

Looking ahead the enzyme volume will be a constant at 10 μL and the DTNB volume will be 60 μL. Then,

$$[ASCh]_{stock} = \frac{M_2V_2}{M_1} = \frac{(largest\ [ASCh]_{assay})(total\ assay\ volume)}{100\ \mu L} \tag{3}$$

4. [ASCh] volume:

Now that you have calculated the [ASCh]$_{stock}$, you can calculate the ASCh volume, V$_1$. This calculation is more complicated than it seems, because the total assay volume, V$_2$, depends on the ASCh added, V$_1$. To simplify the calculation volumes that are constant in each assay can be combined to yield:

$$V_1 = V_{constant} \frac{M_2 - 1}{M_1} \tag{4}$$

Where

$$V_{constant} = V_{buffer} + V_{enzyme} + V_{DNTB} \tag{5}$$

Recall that

> M$_1$ is [ASCh]$_{stock}$ and M$_2$ is [ASCh]$_{assay}$

Thus

$$V_1 = V_{constant} \frac{[ASCh]_{assay} - 1}{[ASCh]_{stock}} \tag{6}$$

5. Selecting an enzyme concentration:

For steady-state kinetics, [S] >>> [E]; that is, the enzyme concentration must be much smaller than the substrate concentration over the entire substrate concentration range. You will start with a prepared [AChE]$_{superstock}$ and dilute a small amount of this 1:100 for use in the kinetic assays. This dilution is [AChE]$_{stock}$. You will then add 10 μl of stock enzyme into each assay mixture. Fill in the Enzyme Volume column in your Table of Volumes for Each Assay.

6. Preparing other reagents:
This AChE assay uses the colorimetric reagent, 5,5´-dithiobis(2-nitrobenzoate) (DTNB). Every assay will contain a constant amount of DTNB. The stock room has prepared $[DTNB]_{stock}$ = 0.01 M. We will use 60 µl of this stock DTNB per assay. Fill in the DTNB Volume column in your Table of Volumes for Each Assay. Finally, complete the Total Assay Volume column in your table.

SUMMARY: An outline for preparing enzyme assays
1. Set the conditions for the assay - choose the assay buffer and approximate assay volume.
2. Select the substrate concentration to use in each assay.
3. Calculate the stock substrate concentration knowing the largest volume and the largest substrate concentration.
4. Calculate the substrate volumes for all other assays knowing the total assay volumes and the stock substrate concentration.
5. Use constant amounts of enzyme and other reagents.

PROCEDURES
Reagents and equipment needs are calculated per six student teams. There is ~20% excess included.

Equipment Needed
  1. Micropipettes, 2-20 μl, 20-200 μl
  2. Volumetric glass pipette, 3.00 ml
  3. UV/vis spectrophotometer
  4. Standard quartz cuvette, 3 ml

Reagents
  1. 0.01 M 5,5′-dithiobis(2-nitrobenzoate) (DTNB) in water. [Prepare 20 ml for 18 groups. Add about 1 mg sodium bicarbonate per ml to help dissolve the DTNB]
  2. 300 units/ml of AChE. [Use 1 mg lyophilized AChE powder (518 units/mg) into 1.7 ml .05 M Na_2HPO_4, pH = 7.0.]

Prepare reagents as follows
  1. Prepare 100 ml of a 1:4 diluted tris buffer from your stock tris buffer.
  2. Prepare 10.00 ml of the stock acetylthiocholine iodide in tris buffer solution.
  3. Prepare 1.0 ml of 1:100 diluted enzyme solution from the superstock AChE solution using your 1:4 diluted tris buffer. Be careful! This is a very expensive reagent. Put your enzyme stock solution on ice.

Assay Procedure
  1 Adjust the spectrophotometer to 412 nm and set the software to record a time course with one spectrum every 15 s for a total of 3 minutes.
  2. Using a volumetric glass pipette, add 3.00 ml of buffer to a quartz cuvette. Then, add a volume of ASCh and mix by gently inverting the cuvette several times. Add 60 μl of the stock DTNB solution and mix. Place the cuvette into the spectrophotometer and start recording the time course. After about 20 seconds, remove the cuvette and add 10 μl of the $[AChE]_{stock}$. Mix rapidly (and gently!) and put the cuvette into the spectrophotometer to complete recording the time course.
  3. Record the initial linear rate as calculated by the spectrophotometer's software (in units of Abs/s).
  4. Repeat this assay procedure for each different [ASCh].

Calculations following completion of assays
  1. Using Beer's law and the molar absorptivity for the colored product, 14,150 $M^{-1}cm^{-1}$, convert all initial linear rates to units of M/min.

2. Using a graphing program, plot velocity (M/min) versus [ASCh] (M) (the Michaelis-Menten plot). If your program can fit a hyperbolic equation (i.e., the Michaelis-Menten equation) to the data, determine K_M and V_{max}.

3. Using a graphing program, construct a Lineweaver-Burk plot (1/v versus 1/[ASCh]). Fit the data to a straight line and calculate both K_M and V_{max}.

4. Calculate the enzyme concentration in the cuvette knowing the superstock [AChE]. [Note: Although the ASCh volume varies from assay to assay, the total assay volume changes little. We can treat the total assay volume as constant and the enzyme concentration as the same for all assays.]

5. Determine the k_{cat} (in units of s^{-1}) and, finally, the measure of catalytic efficiency, k_{cat}/K_M.

Notes to Instructor

This lab assumes a working knowledge of the Michaelis-Menten equation as taught in the biochemistry lecture. Thus, in our schedule, this laboratory often comes near the end of the term.

This lab seeks to emphasize the very practical aspects of steady-state experiments. This means, calculations focus on determining volumes and concentrations. We find it important to emphasize to the students that without these "basic" calculations accurately done, the experiment will fail.

Prelab for AchE Kinetics

1. Briefly outline a step-by-step procedure for this steady-state experiment.

2. List your chosen stock substrate concentration and calculate how you make 10 ml of this solution if the substrate acetylthiocholine iodide has a molar mass of 289.18 g/mol).

3. Construct a table showing the volumes of each reagent you will use for each assay and the corresponding [ASCh]:

Table of Volumes for Each Assay

Assay #	[ASCh]$_{assay}$	Buffer volume	ASCh volume	DTNB volume	Enzyme volume	Total assay volume
1						
2						
etc.						

Lab Report Outline and Point Distribution

1. Introduction
a. Several sentences defining the goal/purpose of this experiment. (3 pts.)
b. Brief outline of the step-by-step procedure for this steady-state experiment. (3 pts.)

2. Data
a. Table of Volumes for Each Assay (the table that you prepared before lab). (3 pts.)
   i. Provide a brief rational for your choice of acetylthiocholine concentrations. (3 pts.)
   i. ii. Give an example of a calculation to determine V_1 for adding ASCh to the cuvette. (3 pts.)
b. Calculations to determine $[AChS]_{stock}$. (4 pts.)
c. Calculations to determine [AChE] in the cuvette. (4 pts.)
d. Table listing [ASCh] versus rates in units of Abs/s and M/min. Give an example of the calculations needed to convert rate from Abs/s to M/min. (4 pts.)

3. Results
a. Computer-generated Michaelis-Menten plot. (5 pts.)
b. Computer-generated Lineweaver-Burk plot. (5 pts.)
c. Values for K_M, V_{max}, k_{cat}, and the ratio k_{cat}/K_M. Show the calculations used to determine k_{cat}. (4 pts.)

4. Analysis
a. Using the ratio, k_{cat}/K_M, comment on the efficiency of acetylcholinesterase. (What does this ratio tell you about the enzyme efficiency?) (5 pts.)
b. Relate the enzyme efficiency to the biological role of acetylcholinesterase. (4 pts.)

Enzyme Kinetics Worksheet

Enzymes do two things really well - specific binding and catalysis:

$$E + S \underset{k_{-1}}{\overset{k_1}{\rightleftharpoons}} ES \overset{k_{cat}}{\longrightarrow} E + P$$

binding   catalysis

The rate of an enzyme-catalyzed reaction is proportional to the concentration of ES:

$$v = k_{cat}[ES] \tag{1}$$

This equation is difficult to use because the [ES] is not directly measurable. An equation that relates the rate to [S] (a variable that can be set by the researcher) is needed. This is called the Michaelis-Menten equation:

$$v = \frac{k_{cat}[E]_{total}[S]}{(K_M + [S])} \tag{2}$$

The Michaelis constant (K_M) combines many rate constants together.

This equation predicts an hyperbola when v is plotted against [S] - just what is normally observed. As you know, this plot approaches a constant v at high [S]. The equation predicts this behavior. When substrate concentration is very large [S]>>(K_M), Equation 2 simplifies to:

$$v = \frac{k_{cat}[E]_{total}[S]}{[S]} = k_{cat}[E]_{total} \tag{3}$$

The rate approaches a constant, large value, $k_{cat}[E]_{total}$. This is termed the maximum velocity or V_{max}. So, the Michaelis-Menten equation can be written:

$$v = \frac{V_{max}[S]}{(K_M + [S])} \tag{4}$$

It can be shown that the K_M is numerically equal to the substrate concentration that gives 1/2 V_{max}.

Michaelis-Menten Plot

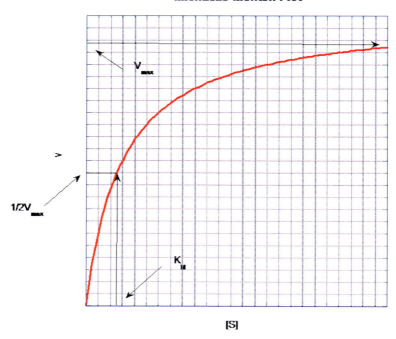

Work the following problems involving the Michaelis-Menten equation and plot.

Problem 1 (3 pts.). Estimate K_M and V_{max} from the following plot:

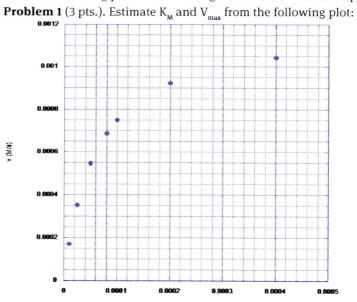

Given $[E]_{total}$ = 5 x 10^{-6} M, calculate the k_{cat} for this enzyme.

Problem 2 (7 pts.). From the following data, prepare a computer-generated Michaelis-Menten plot. Then, estimate V_{max} and K_M.

[S] (M)	v (M/min)
2.0000e-05	6.7961e-05
3.0000e-05	9.2920e-05
4.0000e-05	0.00011382
6.0000e-05	0.00014685
1.0000e-04	0.00019126
0.00020000	0.00024735
0.00030000	0.00027415
0.00050000	0.00030017

Problem 3 (10 pts.) It is common to use a linearized form of the Michaelis-Menten equation to get a more precise measure of K_M and V_{max}. Inverting the Michaelis-Menten equation gives the Lineweaver-Burk equation which is a straight-line equation:

$$\frac{1}{v} = \frac{K_M + [S]}{V_{max}[S]} = \frac{K_M}{V_{max}}\frac{1}{[S]} + \frac{1}{V_{max}}$$

$$Y = (slope)X + (y - intercept)$$

$$(5)$$

Using the above set of data, calculate and plot 1/v versus 1/[S]. Determine K_M and V_{max}. Knowing that the $[E]_{total}$ = 0.00004 M, also calculate k_{cat}.

◊**What can you learn from a steady-state kinetics experiment?**◊

Enzymes do two things really well - specific binding and catalysis. This can be described by following chemical equations:

$$E + S \underset{k_{-1}}{\overset{k_1}{\rightleftharpoons}} ES \xrightarrow{k_{cat}} E + P$$

binding   catalysis

Steady-state kinetics experiments supply numerical measures of an enzyme's capabilities, both catalysis and binding.

<u>Vmax, kcat (the maximum velocity, the catalytic rate constant):</u> In general, V_{max} is not part of published studies because this value depends on $[E]_{total}$. Instead, k_{cat} is reported.

$$V_{max} = k_{cat}[Enzyme]_{total} \qquad (6)$$

The catalytic rate constant (k_{cat}) is a measure of enzyme effectiveness at high [S], when all enzyme sites are saturated with substrate. Substrate is "forced" onto the active site independent of the enzyme's ability to bind. The catalytic rate constant measures one important enzyme characteristic (catalysis) but not the other (specific binding).

Enzymes that are better catalysts have larger k_{cat}'s. This value can be understood as a turnover number. It measures the number of substrates "turned-over" to products per unit time by one enzyme when saturated with substrate. Thus, an enzyme with a k_{cat} = 1500 s^{-1} converts 1500 substrate molecules to products per second under saturating conditions. The other enzyme with a k_{cat} = 4500 s^{-1} is a three-fold more efficient catalyst.

<u>K_M (the Michaelis constant):</u> The Michaelis constant is used in two different ways. First, it gives a good estimate of the "midpoint" for the hyperbolic Michaelis-Menten plot. This follows from the definition of K_M as the [S] that gives 1/2 V_{max}. The K_M allows an estimate of the concentration range over which the enzyme is active.

Second, the Michaelis constant is a ratio of rate constants for the enzyme-catalyzed mechanism. Commonly, the binding step rate constants play a prominent role. So, the K_M can be used as a measure of specific binding - the smaller the K_M the tighter the specific binding. However, it is important to note that the value of K_M is also affected by catalysis (k_{cat}).

<u>k_{cat}/K_M (the ratio of the catalytic rate constant to the Michaelis constant):</u> This ratio is a measure of the overall enzyme efficiency - it takes into account both catalysis (k_{cat}) and a measure of specific binding (K_M). A more efficient enzyme has a larger k_{cat}/K_M while a less efficient enzyme has a smaller ratio. The larger ratio may be caused by the fact that an enzyme is a better catalyst (a larger k_{cat}) or because an enzyme is a better binder (a smaller K_M) or both.

To understand this ratio further we need a more detailed analysis of the reaction mechanism that comes from steady-state theory. Generally:

$$E + S \underset{k_{-1}}{\overset{k_1}{\rightleftharpoons}} ES \overset{k_{cat}}{\longrightarrow} E + P$$

$$\text{binding} \qquad \text{catalysis}$$

As nature improves an enzyme, specific rate constants change. The enzyme can be improved by being made a better catalyst - increasing k_{cat}. And/or, it can become a better enzyme by binding more tightly - decreasing k_{-1}. The third rate constant (k_1) is not affected by the enzyme. Instead, k_1 is a constant and only depends on how fast E and S diffuse together. The k_1 value ranges from about 10^8 $M^{-1}s^{-1}$ to 10^9 $M^{-1}s^{-1}$ under physiological conditions.

For the "perfect" enzyme: (1) ES will almost never fall apart to E plus S (k_{-1} will be very small); ES will almost always immediately react to form products (k_{cat} will be very large). Therfore, substrate will react as quickly as E and S diffuse together. The rate equation that expresses this situation is as follows:

$$E + S \xrightarrow{\ k_1\ } E + P$$

Under these conditions the enzyme can no longer increase the rate of reaction!! The rate depends only on how fast E can diffuse together with S. The reaction rate is out of the enzyme's control!

So, there is an upper limit to the rate constant of an enzyme-catalyzed reaction (10^8 $M^{-1}s^{-1}$ to 10^9 $M^{-1}s^{-1}$). Diffusion serves as nature's speed limit. As enzymes improve, their rate constants approach this limit. More specifically, k_{cat}/K_M approaches this limit. That is, the closer k_{cat}/K_M is to 10^8 $M^{-1}s^{-1}$ to 10^9 $M^{-1}s^{-1}$, the closer the enzyme is to "perfection."

[A brief mathematical justification for using k_{cat}/K_M to judge enzyme perfection.]

Under steady-state conditions,

$$K_M = (k_{-1} + k_{cat})/k_1 \tag{7}$$

for the common mechanism,

$$E + S \underset{k_{-1}}{\overset{k_1}{\rightleftarrows}} ES \xrightarrow{\ k_{cat}\ } E + P$$

$$\text{binding} \qquad \text{catalysis}$$

The enzyme can be improved by becoming a better catalyst, increasing k_{cat} and/or it can become a better enzyme by binding more tightly, decreasing k_{-1}. Either way, $(k_{-1} + k_{cat})$ approaches k_{cat} for the best enzymes and Equation (7) becomes:

$$K_M = k_{cat}/k_1 \tag{8}$$

Then, the ratio k_{cat}/K_M equals $k_{cat}/(k_{cat}/k_1)$ which is just equal to k_1. The efficiency of the enzyme reaches an upper limit, k_1. No matter how much better the enzyme catalyzes the reaction (increasing k_{cat}) or how much more tightly it binds substrate (decreasing k_1), the overall reaction rate will not improve further. At this point, the reaction rate just depends on the speed at which S and E diffuse together. The enzyme is doing all it can and it is said to have reached "perfection."

The ratio, k_{cat}/K_M, is perhaps the most useful constant for describing enzyme efficiency. First, it allows a relative comparison between enzymes - a comparison that takes into account both catalysis and binding. Second, this ratio provides an absolute comparison to the "perfect" enzyme.

Problem 4 (20 pts.). Two enzymes that catalyze the same reaction were studied ([Enzyme A] = 5 x 10^{-7} M; [Enzyme B = 2 x 10^{-6} M). The data from this study are given below:

[S] (M)	v (M/s), Enz A	v (M/s), Enz B
2.0000e-05	3.2500e-05	0.0018824
3.0000e-05	4.7273e-05	0.0021818
4.0000e-05	6.1176e-05	0.0023704
6.0000e-05	8.6667e-05	0.0025946
1.0000e-04	0.00013000	0.0028070
0.00020000	0.00020800	0.0029907
0.00030000	0.00026000	0.0030573
0.00050000	0.00032500	0.0031128

(1) Create Lineweaver-Burk plots for each enzyme.
(2) Calculate and report K_M, V_{max}, k_{cat} and k_{cat}/K_M (show calculations as well as reporting values).
(3) Compare the enzymes with each other and with "perfection."

14 Separation of the Phosphatidylcholines Using Reverse Phase HPLC

14.1 Learning Objective

This laboratory has 2 goals, (1) to learn more of membrane lipid structures by working with phosphatidylcholines and (2) to learn the basics of an especially important high performance liquid chromatography (HPLC) technique, reverse phase HPLC. You should use your knowledge of phosphatidylcholine structures to rationalize the elution pattern from the HPLC.

14.2 Phosphatidylcholines

Phosphatidylcholine is an important class of lipids (hydrophobic biochemicals). This class is one of the primary constituents of the biological membrane. Phosphatidylcholines have a common structure. To build a phosphatidylcholine, start with glycerol. Other components are connected to glycerol using ester bonds. Two fatty acids (long chain carboxylic acids) are esterified to the top two positions of glycerol. The third position contains a phosphate and a choline. Each phosphatidylcholine differs from others of its class based on the molecular characteristics of the fatty acid chains (*Fig. 14.1*).

14.3 High Performance Liquid Chromatography (HPLC)

Separation of phosphatidylcholines is difficult, but can be done using high performance liquid chromatography (HPLC). This separation technique depends on passing a solution (the mobile phase) through a column packed with very tiny particles (the stationary phase). Some solutes are attracted strongly to these particles, and travel through the column slowly. These solutes stick to a particle for a certain length of time and then "hop" to the next particle. Compared with the motion of the mobile phase, these solutes are retarded. Other solutes are only attracted to the particles weakly and so can travel through the column quickly.

The differential movement of solutes leads to the separation of the solute. This is often shown as a chromatogram (*Fig. 14.2*).

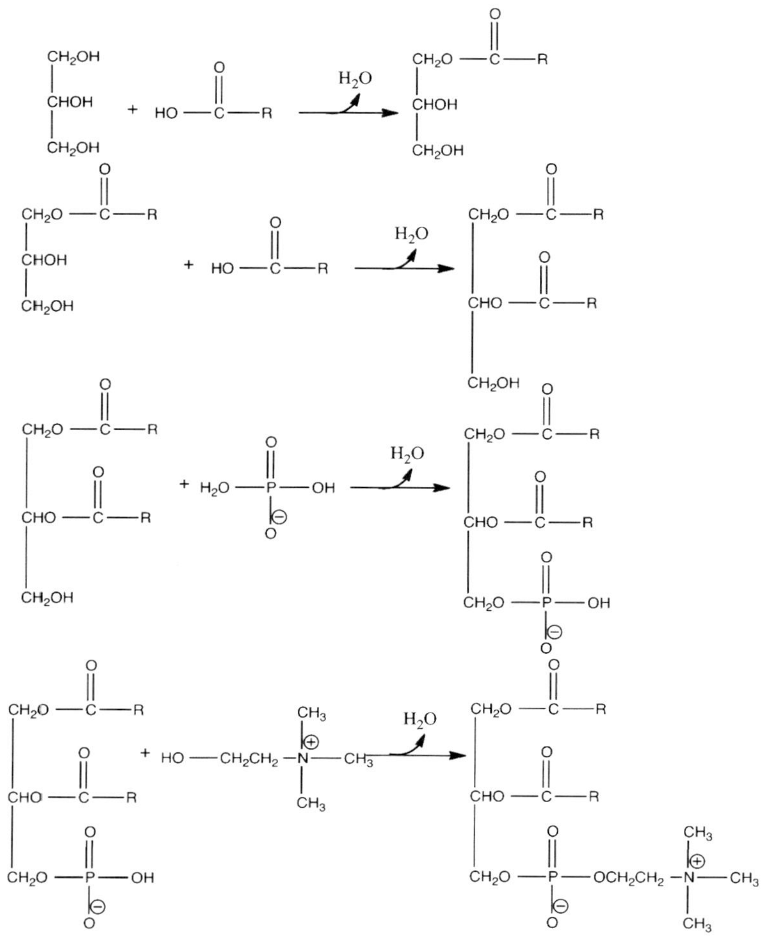

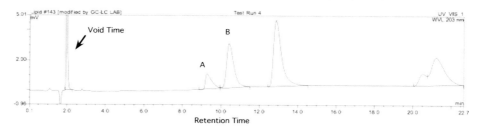

Figure 14.1: Schematic showing of the connections needed to build a phosphatidylcholine.

Figure 14.2: A chromatogram showing the separation of molecule A from molecule B: The x-axis is measured as retention time, the time a solute spends on the chromatography column after the sample is injected. The void time is the fastest time a solute pass through the column, i.e., the solute does not stick to the stationary phase at all. "(Reproduced with permission, from M. Ferrer, O.V. Golyshina, F.J. Plou, K.N. Times and P.N. Golyshin, (2005), *Biochemical Journal*, 391(2) 269-276. © the Biochemical Society)

You will be using *reverse phase HPLC*. The particles in this technique are made of silica (sand) that has been coated with alkane chains. Solutes that are more hydrophobic are more strongly attracted to the stationary phase and move more slowly through the column. You will be separating five phosphatidylcholine compounds with very similar structures (*Fig. 14.3*).

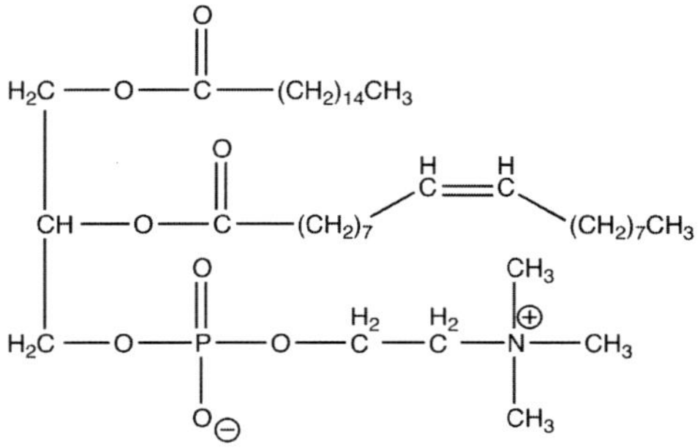

1-palmitoyl-2-oleoylglycerophosphocholine
(POPC)

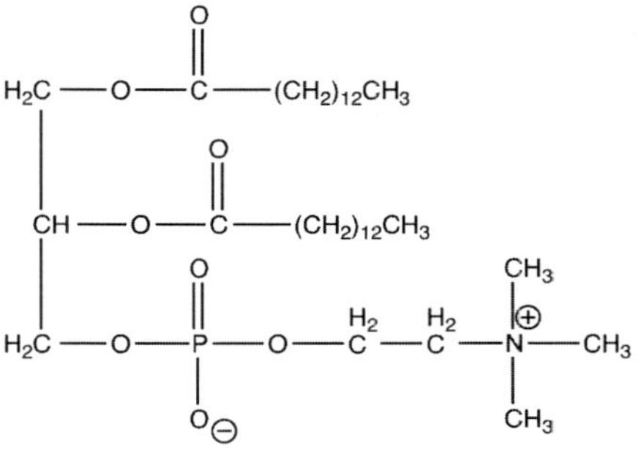

1,2-dimyristoylglycerophosphocholine
(DMPC)

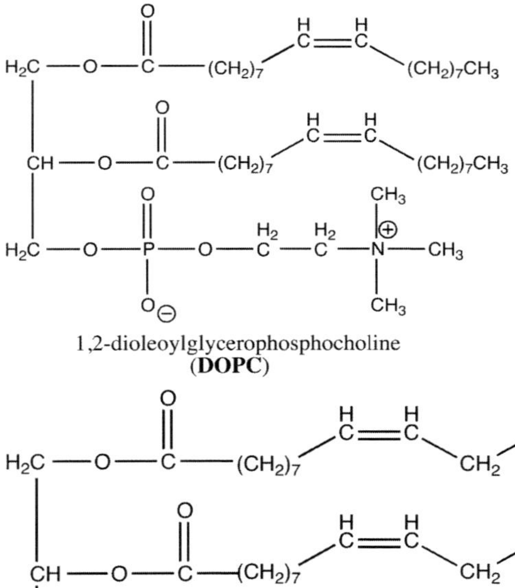

1,2-dioleoylglycerophosphocholine
(DOPC)

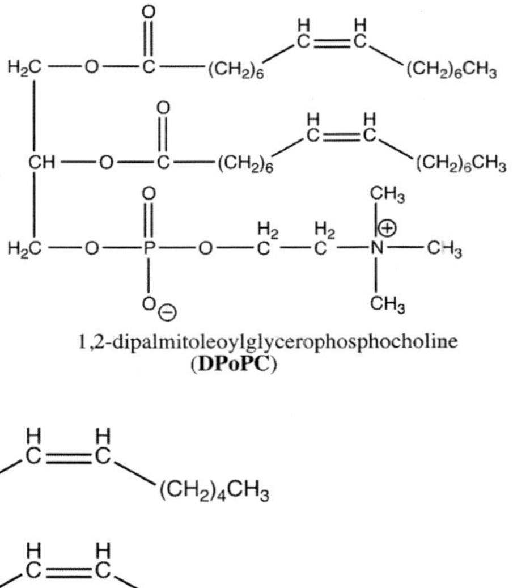

1,2-dipalmitoleoylglycerophosphocholine
(DPoPC)

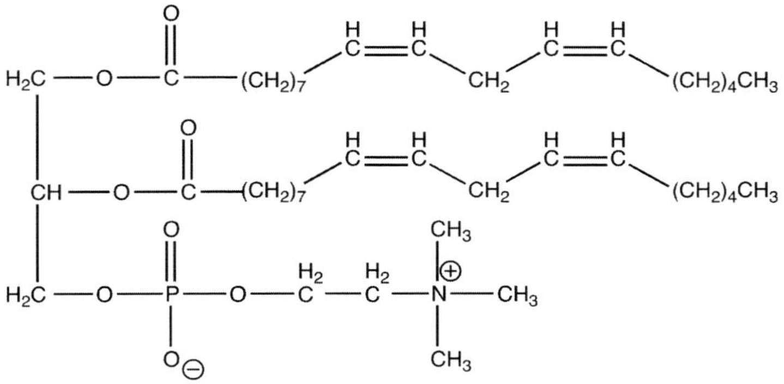

1,2-dilinoleoylglycerophosphocholine
(DLPC)

Figure 14.3: Structures of the phosphatidylcholines to be separated in the experiment.

Phosphatidylcholines with the longer fatty acid chains are more strongly attracted to the stationary phase. The length of the fatty acid chains depends on:
- the number of carbons,
- the presence of cis double bonds. Each cis double bond makes the chain act as if it were one to two carbons shorter (less hydrophobic).

Which of the five phosphatidylcholines would you predict to be attracted most to the stationary phase? Which would you predict will be attracted least? Based on the PC structure, predict the order of elution from the column.

14.4 Quantifying Chromatography

The success of a separation can be measured in several different ways. First, quicker separation tends to be better, because the experimenter doesn't have to wait for her/his results too long. For each solute a retention time is measured. This is the elapsed time from the beginning to when the solute peak leaves the column. Typically, this time is reported relative to the quickest elution time (the time it takes for solvent to pass through the column, the void time). So, the quickness of separation is measured by relative retention or capacity factor (k').

$$k' = \frac{(t_{peak} - t_{void})}{t_{void}} \tag{1}$$

Using the chromatogram in Fig. 14.2, Peak A has a retention time of 9.3 min. The void time for this chromatogram is 2.0 min.

$$k' = \frac{(9.3 \text{ min} - 2.0 \text{ min})}{2.0 \text{ min}} = 3.6$$

Peak A is retained on the column 3.6-times longer than the mobile phase.

Another measure of success is to monitor the shape of each peak. Sharp peaks mean a better separation. This is measured by determining the width of the peak relative to how long the peak is retained on the column.

$$N = \text{column efficiency} = 16 \left(\frac{\text{retention peak time}}{\text{peak width}} \right)^2 \tag{2}$$

The Peak A from Fig. 14.2 has a retention time of 9.3 min with a peak width of 0.8 min.

$$N = 16 \left(\frac{9.3 \text{ min}}{0.8 \text{ min}} \right)^2 = 2162$$

A well-done separation will give efficiencies in the thousands.

A third measure of success is to quantify how much separation occurs between neighboring peaks. This selectivity is calculated as the ratio of capacity factors

$$\alpha = \text{selectivity} = \frac{k_2^{'}}{k_1^{'}} \tag{3}$$

Peak A has a $k^{'} = 3.6$ while Peak B has a $k^{'} = 4.3$. This means, this separation between Peak A and Peak B has a selectivity of 1.2. Peak B is retained 20% longer than Peak A. A good separation will give selectivity values of greater than 1.1.

PROCEDURES

Reagents and equipment needs are calculated per six student teams. There is ~20% excess included.

Equipment/glassware needed:
1. Standard HPLC system 1 per 2 student teams
2. C-18 reverse phase HPLC column

Reagents needed:
1. 98% methanol
2. 100 μl of phosphatidylcholines mix dissolved in methanol. Stock concentrations for each phosphatidylcholine in the mix are listed below
    a. DMPC 10 mg/ml
    b. DPoPC 5 mg/ml
    c. DLPC 1 mg/ml
    d. POPC 5 mg/ml
    e. DOPC 5 mg/ml

Experimental procedure:
1. A standard analytical, reverse phase HPLC column (C-18) is equilibrated with 98% methanol - 2% water. A flow rate of 1 ml/min is convenient.
2. Each separation uses 10 μl of a mixed phosphatidylcholine sample.
3. The sample contains 10 mg/ml DMPC, 5 mg/ml DPoPC, 1 mg/ml DLPC, 5 mg/ml POPC, 5 mg/ml DOPC in methanol.
4. Each separation requires about forty minutes.
5. Follow your instructor's directions concerning operation of the HPLC chromatograph.

Data Analysis:
1. Calculate the capacity factor (relative retention) for each phosphatidylcholine.
2. Determine the column efficiency (N) calculated using the DLPC peak.
3. Calculate the selectivity (α) between (a) DMPC vs. DPoPC, (b) DPoPC vs. DLPC, (c) DLPC vs. POPC and (d) POPC vs. DOPC.

Notes to Instructor

This laboratory is scheduled to maximize use of a limited number of chromatographs. At the authors' institution, we use three HPLC machines simultaneously. Two student teams (each team composed of a student pair) are assigned to each chromatograph. While one team runs the chromatography, the other team is completing an in-lab HPLC problem set. Thus, by the end of the period, all teams have completed a chromatographic trial and practiced the common calculations needed to analyze a chromatogram.

HPLC of Lipids Prelab

1. Draw the structure of each phosphatidyl choline that you are going to separate during lab (there are five)!

2. Circle the hydrophobic part of each molecule!

3. Rank these molecules based on hydrophobicity from least (5) to most hydrophobic (1)!

4. Which of the five PCs would you predict to be attracted most to the stationary phase? Which would you predict will be attracted least?

5. Based on the PC structure, predict the order of elution from the column.

HPLC of Phosphatidylcholines
Lab Report Outline and Point Distribution

Introduction
1. Several sentences defining the goal/purpose of this experiment. (3 pts.)

Data
1. A copy of your chromatogram with each peak labeled with a specific phosphatidylcholine. (10 pts.)

Results (please show all calculations)
1. The capacity factor (relative retention) for each phosphatidylcholine. (10 pts.)
2. The column efficiency (N) calculated using the DLPC peak. (4 pts.)
3. The selectivity (α) between (a) DMPC vs. DPoPC, (b) DPoPC vs. DLPC, (c) DLPC vs. POPC, and (d) POPC vs. DOPC. (8 pts.)

Analysis
1. Which phosphatidylcholines are cleanly separable on this column. Briefly explain. (5 pts.)
2. Problems (10 pts.)

HPLC Problem Set

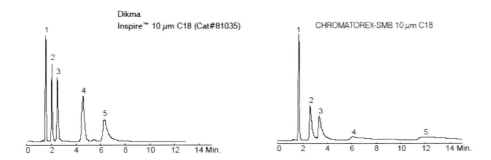

(Courtesy of Dikma Technologies. Chromatorex is a registered trademark of Fuji Silysia Chemical Ltd. Dikma Technologies Inc. is not affiliated with the above company.

1. (5 pts.) The chromatogram from Chromatorex-SMB appears not as good as the chromatogram from Inspire. Calculate the column efficiencies (N) based on peak 3. (Use a ruler and the conversion, 1 minute/6 mm.) Does this agree with the conclusion in the first sentence? Briefly explain.

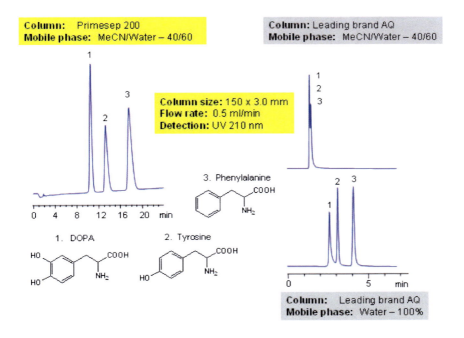

(Courtesy of SIELC Technologies)

2. (5 pts.)By just looking at the chromatograms, rank them from best separation to worst separation. Using the two chromatograms that are measurable, determine retention times for each peak. (Estimate times to the nearest 0.1 minute.) Calculate capacity factors for each peak using 1.1 min as the void time. Then, calculate selectivity factors (α) for Peak 1 vs. Peak 2 and for Peak 2 vs. Peak 3. Do the selectivity factors agree with your ranking? Briefly explain.

List of Figures